HILBERT'S
FOURTH PROBLEM

SCRIPTA SERIES IN MATHEMATICS

Tikhonov and Arsenin • Solutions of Ill-Posed Problems, 1977

Rozanov • Innovation Processes, 1977

Pogorelov • The Minkowski Multidimensional Problem, 1978

Kolchin, Sevast'yanov, and Chistyakov • Random Allocations, 1978

Boltianskiĭ • Hilbert's Third Problem, 1978

Besov, Il'in, and Nikol'skiĭ • Integral Representations of Functions and Imbedding Theorems, Volume I, 1978

Besov, Il'in, and Nikol'skii • Integral Representations of Functions and Imbedding Theorems, Volume II, 1978

Sprindžuk • The Metric Theory of Diophantine Approximations, 1979

Krushkal' • Quasiconformal Mappings and Riemann Surfaces, 1979

Mirkin • Group Choice, 1979

Pogorelov • Hilbert's Fourth Problem, 1979

HILBERT'S FOURTH PROBLEM

Aleksei Vasil'evich Pogorelov
USSR Academy of Sciences

Translated by
Richard A. Silverman

Edited by
Irwin Kra
SUNY at Stony Brook

in cooperation with
Eugene Zaustinskiy
SUNY at Stony Brook

1979

V. H. WINSTON & SONS
Washington, D.C.

A HALSTED PRESS BOOK

JOHN WILEY & SONS

New York   Toronto   London   Sydney

V. H. Winston & Sons, a Division of Scripta Technica, Inc.,
Publishers
1511 K Street, N.W., Washington, D.C. 20005

Distributed solely by Halsted Press, a Division of John Wiley
& Sons, Inc.

Library of Congress Cataloging in Publication Data:

Pogorelov, Alekseĭ Vasil'evich, 1919–
    Hilbert's fourth problem.

    (Scripta series in mathematics)
    Translation of Chetvertaía problema Gil'berta.
    Bibliography: p.
    1. Geometry–Foundations. I. Kra, Irwin.
II. Zaustinskiy, Eugene. III. Title. IV. Se-
ries.
QA681.P5913   516'.1   79–14508
ISBN 0–470–26735–6 (Halsted)

Composition by Isabelle K. Sneeringer, Scripta Technica, Inc.

CONTENTS

PREFACE TO THE AMERICAN EDITION 1

TRANSLATION EDITOR'S FOREWORD 3

INTRODUCTION 5

SECTIONS
    1. Projective Space 9
    2. Projective Transformations 13
    3. Desarguesian Metrizations of Projective Space 19
    4. Regular Desarguesian Metrics in the
       Two-Dimensional Case 24
    5. Averaging Desarguesian Metrics 31
    6. The Regular Approximation of Desarguesian Metrics 38
    7. General Desarguesian Metrics in the
       Two-Dimensional Case 46
    8. Funk's Problem 54
    9. Desarguesian Metrics in the Three-Dimensional Case ... 61

10. Axioms for the Classical Geometries 68
11. Statement of Hilbert's Problem 75
12. Solution of Hilbert's Problem 82

NOTES 88

BIBLIOGRAPHY 93

INDEX 95

PREFACE TO THE
AMERICAN EDITION

Hilbert's Fourth Problem, as presented by Hilbert himself (see the Introduction), is stated in rather broad and general terms. In brief, the problem consists in the investigation of metric spaces which admit a geodesic mapping onto a projective space or a domain of such a space.[1*] Hilbert's Fourth Problem is related to the foundations of geometry, the calculus of variations, and differential geometry.[2] We shall consider Hilbert's Problem as a problem in the foundations of geometry and, in this regard, following Hilbert, we formulate the problem more precisely as follows.

Suppose we take the system of Axioms for Euclidean geometry, drop those axioms involving the concept of angle, and then supplement the resulting system with the "triangle inequality," regarded as an axiom. The resulting system of axioms is incomplete and there exist infinitely many geometries, in addition to Euclidean geometry, which satisfy these axioms. Hilbert's Problem consists in describing all

*Note. Superscripts refer to Notes beginning on page 88.

1

possible geometries satisfying this system of axioms. The present work is devoted to the solution of the problem as stated in this form. The problem will be considered from the standpoint of all three classical geometries, namely those of Euclid, Lobachevski and Riemann.

It turns out that the solution of the problem, as formulated here, reduces to the determination of all of the so-called Desarguesian metrics[3] in projective space; that is, metrics for which the geodesics are straight lines. Such metrics were obtained by Hamel (see [12]), under the assumption of sufficient smoothness. However, a complete solution of the problem requires the determination of all Desarguesian metrics, without assuming smoothness, subject only to the condition of continuity which is guaranteed by the axioms.

The occasion for the present investigation is a remarkable idea due to Herbert Busemann, which I learned about from his report to the International Congress of Mathematicians at Moscow in 1966. Busemann gave an extremely simple and very general method of constructing Desarguesian metrics by using a nonnegative completely additive set function on the sets of planes and defining the length of a segment as the value of this function on the set of planes intersecting the segment.

I suspected that all continuous Desarguesian metrics could be obtained by this method. The proof of this in the two-dimensional case strengthened my belief in this conjecture and I announced a general theorem in [18]. However, it turned out later, on making a detailed investigation of the three-dimensional case, that the completely additive set function figuring in Busemann's construction may not satisfy the condition of nonnegativity. Therefore, the result given here, while preserving its original form, assumes that other conditions are satisfied.

This book is addressed to a wide circle of readers and, accordingly, it begins with a review of the basic facts of the geometry of projective space (Sections 1, 2). A description of the axiom systems of the classical geometries is given in Section 10. A detailed exposition of the core of the problem, as formulated here, is given in Section 11, together with illustrative examples.

I regard it as my pleasant duty to thank the publisher V. H. Winston & Sons for the interest shown in my work.

A. V. Pogorelov

TRANSLATION EDITOR'S FOREWORD

It has been slightly more than seventy-five years since David Hilbert presented a list of twenty-three outstanding and important problems to the Second International Congress of Mathematicians held in Paris in 1900. Surprisingly, very few books have appeared about this list of problems; despite the tremendous progress during the last three quarters of a century toward the solutions of practically the entire list of problems.

Hilbert's fourth problem (find all geometries in which the "ordinary lines" are the "geodesics") is particularly attractive. The problem is elementary enough that it can certainly be understood and appreciated by a beginning graduate student of mathematics. However, its solution though of a generally elementary character brings together ideas and tools from many diverse and interesting branches of mathematics: geometry, analysis (especially ordinary and partial differential equations), and the calculus of variations.

A partial solution, under strong assumptions, to Hilbert's fourth problem was already obtained by Georg Hamel in 1901. This present work by A. V. Pogorelov originated from a remarkable idea of Herbert Busemann (in his report to the International Congress of Mathematicians of Moscow in 1966). Pogorelov slightly reformulates Hilbert's problem, and proceeds on the basis of this new idea to give an extremely elegant solution—a real mathematical gem.

This book is extremely well written. Most of the prerequisites (with the exception of standard portions of advanced calculus) are developed as needed. The reader who studies this volume will not only discover how one particular problem is solved, but will also pick up a lot of interesting mathematics on the way.

The English translation was reviewed by Eugene Zaustinsky, who also supplied a very useful set of notes that guide the reader to more literature on the subject.

Pogorelov's book is a welcome addition to the mathematical literature. It will especially be appreciated by those interested in geometry and the foundations of geometry.

Irwin Kra

INTRODUCTION

In the year 1900, at the Second International Congress of Mathematicians in Paris, David Hilbert formulated a number of problems whose investigation would, in his opinion, greatly stimulate the further development of mathematics. His fourth problem was devoted to the foundations of geometry, and consists of the following, as stated by Hilbert himself ([13], pp. 449-451):

"If, from among the axioms necessary to establish ordinary euclidean geometry, we exclude the axiom of parallels, or assume it as not satisfied, but retain all other axioms, we obtain, as is well-known, the geometry of Lobachevski (hyperbolic geometry). We may therefore say that this is a geometry standing next to euclidean geometry. If we require further that that axiom be not satisfied whereby, of three points of a straight line, one and only one lies between the other two, we obtain Riemann's (elliptic) geometry, so that this geometry appears to be the next after Lobachevsky's. If we wish to carry out a similar investigation with respect to the axiom of Archimedes, we must look upon this as not satisfied, and we arrive

5

thereby at the non-archimedean geometries which have been investigated by Veronese and myself [14]. A more general question now arises: Whether from other suggestive standpoints geometries may not be devised which, with equal right, stand next to euclidean geometry. Here I should like to direct your attention to a theorem which has, indeed, been employed by many authors as a definition of a straight line, viz., that the straight line is the shortest distance between two points. The essential content of this statement reduces to the theorem of Euclid that in a triangle the sum of two sides is always greater than the third side—a theorem which, as easily seen, deals solely with elementary concepts, i.e., with such as are derived directly from the axioms, and is therefore more accessible to logical investigation. Euclid proved this theorem, with the help of the theorem of the exterior angle, on the basis of the congruence theorems. Now it is readily shown that this theorem of Euclid cannot be proved solely on the basis of those congruence theorems which relate to the application of segments and angles, but that one of the theorems on the congruence of triangles is necessary. We are asking, then, for a geometry in which all the axioms of ordinary euclidean goemetry hold, and in particular all the congruence axioms except the one of the congruence of triangles (or all except the theorem of the equality of the base angles in the isosceles triangle), and in which, besides, the proposition that in every triangle the sum of two sides is greater than the third is assumed as a particular axiom.

One finds that such a geometry really exists and is none other than that which Minkowski constructed in his book, Geometrie der Zahlen [16] and made the basis of his arithmetical investigations. Minkowski's Geometry is therefore also a geometry standing next to the ordinary euclidean geometry; it is essentially characterized by the following stipulations:

1. The points which are at equal distances from a fixed point O lie on a convex closed surface of the ordinary euclidean space with O as a center.

2. Two segments are said to be equal when one can be carried into the other by a translation of the ordinary euclidean space.

In Minkowski's geometry the axiom of parallels also holds. By studying the theorem of the straight line as the shortest distance between two points, I arrived ([15] and [14], Appendix I) at a geometry in which the parallel axiom does not hold, while all other axioms of Minkowski's geometry are satisfied. The theorem of the straight line as the shortest distance between two points and the essentially equivalent theorem of Euclid about the sides of a triangle, play an important part not only in number theory but also in the theory of surfaces and the calculus of variations. For this reason, and because I believe that the thorough investigation of the conditions for the validity of this theorem will throw a new light upon the idea of distance, as well as upon other elementary ideas, *e.g.,* upon the idea of the plane, and the possibility of its definition by means of the idea of the straight line, *the construction and systematic treatment of the geometries here possible seem to me desirable.*

In the case of the plane and under the assumption of the continuity axiom, the indicated problem leads to the question treated by Darboux ([10] p. 59): Find all variational problems in the plane for which the solutions are all the straight lines of the plane—a question which seems to me capable and worthy of far-reaching generalizations [16]."

This book is devoted to Hilbert's fourth problem [13] and contains its solution when formulated as follows: Find to within an isomorphism all realizations of the axiom systems of the classical geometries (Euclidean, Lobachevskian and elliptic) if, in these systems, we drop the axioms of congruence involving the concept of angle and supplement the systems with the "triangle inequality," regarded as an axiom.

The first and, indeed, the only work devoted to Hilbert's problem in this formulation is due to Hamel [12] the other works being devoted to the study of special Desarguesian spaces. Hamel showed that every solution of Hilbert's problem can be represented in a projective space, or a convex domain of such a space, if congruence of segments is defined as equality of their lengths in a special metric, for which the lines of the space are geodesics. (Such metrics are called Desarguesian metrics.) Thus, the solution of Hilbert's problem was reduced to the problem of the constructive definition of all Desarguesian metrics. Hamel solved this problem under the assumption of a sufficiently regular metric. However, as simple examples show, regular plane metrics by no means exhaust the class of all plane metrics, and the axioms of the geometries under consideration imply only continuity of the metrics. Therefore, a complete solution of Hilbert's problem entails a constructive definition of all continuous Desarguesian metrics and this is the problem to which the present work is devoted.

A. V. Pogorelov

§1. PROJECTIVE SPACE

By a *point* of projective space we mean an ordered quadruple of real numbers $x = (x_1, x_2, x_3, x_4)$, which are not all zero. Proportional quadruples are regarded as *equivalent,* and define the same point of space. The numbers x_1, x_2, x_3, x_4 are called *homogeneous coordinates.*[4]

By a *plane* we mean the set of points satisfying a linear equation

$$a_1x_1 + a_2x_2 + a_3x_3 + a_4x_4 = 0,$$

and by a *line* we mean the intersection of two distinct planes. Thus, a line is specified by a system of two equations

$$\begin{cases} a_1x_1 + a_2x_2 + a_3x_3 + a_4x_4 = 0, \\ a_1'x_1 + a_2'x_2 + a_3'x_3 + a_4'x_4 = 0, \end{cases}$$

where the rank of the matrix

$$\begin{pmatrix} a_1 & a_2 & a_3 & a_4 \\ a_1' & a_2' & a_3' & a_4' \end{pmatrix}$$

equals two. It is convenient to use vector notation for the equations of lines and planes. We set

$$a \cdot x = a_1x_1 + a_2x_2 + a_3x_3 + a_4x_4,$$

and can then write the equation of a plane in the form $a \cdot x = 0$, and the equation of a line in the form $a \cdot x = 0, a' \cdot x = 0$, where a and a' are linearly independent vectors. We now note some properties of lines and planes.

Let x' and x'' be two distinct points of a line. *Then every point of the line has a representation $x = \lambda'x' + \lambda''x''$, where λ' and λ'' are real numbers, not both zero.* Conversely, every point with this representation belongs to the line. In fact, the line is specified by a system of equations

$$a' \cdot x = 0, \quad a'' \cdot x = 0 \tag{1}$$

and the points x' and x'' satisfy this system. Since the rank of the system is two, every solution x of the system is a linear combination of the independent solutions x' and x'', i.e., $x = \lambda'x' + \lambda''x''$. The fact that every point x with this representation satisfies the system (1) is obvious. In view of the indicated representation of a point x of the line, in terms of two given distinct points x' and x'', we conclude that a line is uniquely determined by any two of its distinct points. Hence, *no more than one line passes through two distinct points.*

We next show that *there is a line passing through any two distinct points.* Let the given points be x' and x'' and consider the system of equations

$$a \cdot x' = 0, \quad a \cdot x'' = 0 \tag{2}$$

in a. Since the points x' and x'' are distinct, the rank of the matrix

$$\begin{pmatrix} x_1' & x_2' & x_3' & x_4' \\ x_1'' & x_2'' & x_3'' & x_4'' \end{pmatrix}$$

of the system is two. Therefore the system has two independent solutions a' and a'', and the line specified by the equations

$$a' \cdot x = 0, \quad a'' \cdot x = 0$$

passes through the points x' and x''. Q.E.D.

On every line there are two distinct points. In fact, let $a' \cdot x = 0$, $a'' \cdot x = 0$ be the equations of the line. This system of equations $a' \cdot x = 0$, $a'' \cdot x = 0$ in x has two linearly independent solutions x' and x'', since the rank of the system is two. These solutions give two distinct points of the line determined by the intersection of the planes $a' \cdot x = 0$, $a'' \cdot x = 0$. Q.E.D.

Three distinct points x', x'', x''' *lie on the same line if and only if the rank of the matrix*

$$\begin{pmatrix} x_1' & x_2' & x_3' & x_4' \\ x_1'' & x_2'' & x_3'' & x_4'' \\ x_1''' & x_2''' & x_3''' & x_4''' \end{pmatrix}$$

is two. In fact, if the points x' and x'' are distinct, then $x''' = \lambda' x' + \lambda'' x''$, by what has been proven. The rank of the matrix is less than three because its rows are linearly dependent. Conversely, if the rank of the matrix is less than three, it must be two since the points x' and x'' are distinct. But, then, the third row can be expressed as a linear combination of the first and second rows; i.e., $x''' = \lambda' x' + \lambda'' x''$. This means that the point x''' lies on the line passing through the points x' and x''. Q.E.D.

There is one and only one plane passing through three non-collinear points. In fact, let x', x'', x''' be the given points and consider the system of equations

$$a \cdot x' = 0, \qquad a \cdot x'' = 0, \qquad a \cdot x''' = 0 \qquad (3)$$

in a. The rank of the system is three since the points x', x'', and x''' are noncollinear. The system (3), therefore, has a nontrivial solution a, which is uniquely determined up to a nonzero factor. The plane $a \cdot x = 0$ passes through the given points and is unique, by the uniqueness of the solution of the system (3). Q.E.D.

There is one and only one plane passing through a line and a point not lying on the line. Let us mark two distinct points on the line and draw a plane through them and the given point. This plane contains the given line and passes through the given point. Every plane passing through the given line contains the two distinct points which we marked on it. Our plane is unique because three distinct noncollinear points of a plane determine the plane uniquely. Q.E.D.

A line that does not lie in a plane intersects the plane in one and only one point. Let $a' \cdot x = 0$ be the equation of the plane and let $a'' \cdot x = 0$ and $a''' \cdot x = 0$ be the equations of the line. The system of homogeneous equations $a' \cdot x = 0$, $a'' \cdot x = 0$, $a''' \cdot x = 0$ always has a nontrivial solution x, and this solution gives a point lying on the intersection of the line and the plane. If there were two distinct such points, then, by what has been proven, the line would lie in the plane, contrary to hypothesis. Q.E.D.

Two distinct lines, lying in the same plane, intersect in one and only one point. Let α be the plane and let g_1 and g_2 be the distinct lines lying in the plane α. Take a point x that does not lie in the plane α, and draw planes α_1 and α_2 through x and the lines g_1 and g_2, respectively. The planes α, α_1, α_2 intersect and the point of intersection belongs to the lines g_1 and g_2. Since the lines g_1 and g_2 are distinct, they cannot have other points of intersection, by what has already been proven. Q.E.D.

Let x and y be two points on a line. Then, every point of the line, other than x and y, has a representation $\lambda x + \mu y$, $\lambda\mu \neq 0$. A set of points of the line, for which $\lambda\mu$ has a fixed sign, is called a (*projective line*) *interval with endpoints x and y*. On a projective line there are two intervals with endpoints x and y. We have $\lambda\mu > 0$ on one of these intervals and $\lambda\mu < 0$ on the other. This definition of line interval is obviously independent of the normalization of the coordinates of the endpoints x and y.

Let x^1, x^2, x^3 be three noncollinear points. The figure consisting of these points and three intervals joining them in pairs will be called a *projective triangle* if there exists a plane which does not intersect the triangle.[5] The points x^1, x^2, x^3 are called the *vertices* of the triangle and the intervals joining them are called the *sides* of the triangle.

A projective triangle satisfies the *Axiom of Pasch*: *If a plane does not pass through the vertices of a triangle and intersects one of its sides, then it intersects one and only one of the other two sides of the triangle.* The proof is as follows. Let $a \cdot x = 0$ be the equation of a plane which does not intersect the triangle. Then $a \cdot x^i \neq 0$, $i = 1, 2, 3$, since this plane does not pass through the vertices of the triangle. We may assume, without loss of generality, that $a \cdot x^i > 0$, as this can always be achieved through a suitable normalization of the points x^i. Because $a \cdot x^k > 0$, the points of the side of the triangle joining the vertices x^i and x^j have a representation $u = \lambda x^i + \mu x^j$, for which $\lambda\mu > 0$. In fact, if we had $\lambda\mu < 0$, we could find values of λ and μ for which $a \cdot u = \lambda a \cdot x^i + \mu a \cdot x^j = 0$; that is, the side $x^i x^j$ of the triangle would intersect the plane $a \cdot x = 0$, contrary to hypothesis.

Next, let $b \cdot x = 0$ be the equation of a plane which intersects the side $x^1 x^2$ of the triangle, but does not pass through any of its vertices. Since λ and μ have the same sign on the side $x^1 x^2$ of the triangle and since this side intersects the plane $b \cdot x = 0$, the quantities $b \cdot x^1$ and $b \cdot x^2$ are both different from zero and have opposite signs. We may suppose, without loss of generality, that $b \cdot x^1 > 0$ and $b \cdot x^2 < 0$. We know that $b \cdot x^3 \neq 0$. If we assume that $b \cdot x^3 > 0$, then the side $x^1 x^3$ of the triangle does not intersect the plane $b \cdot x = 0$, since $b \cdot (\lambda x^1 + \mu x^3) = \lambda b \cdot x^1 + \mu b \cdot x^3 > 0$. On the other hand, the side $x^2 x^3$ does intersect this plane since, if $b \cdot x^2 < 0$ and $b \cdot x^3 > 0$, we can find $\lambda, \mu > 0$ for which $b \cdot (\lambda x^2 + \mu x^3) = \lambda b \cdot x^2 + \mu b \cdot x^3 = 0$. Similarly, if we assume that $b \cdot x^3 < 0$, then the side $x^1 x^3$ intersects the plane $b \cdot x = 0$ and the side $x^2 x^3$ does not. Thus, in any case, the plane $b \cdot x = 0$, which intersects the side $x^1 x^2$, intersects either the side $x^1 x^3$ or the side $x^2 x^3$.

This completes the proof of the Axiom of Pasch.

§2. PROJECTIVE TRANSFORMATIONS

A one-to-one mapping of projective space onto itself is said to be *projective* if it carries planes into planes and, therefore, lines into lines. *A linear transformation,* that is, *a transformation given by the formulas*

$$x'_i = \sum_\alpha a_{i\alpha} x_\alpha \qquad (i = 1, 2, 3, 4) \tag{1}$$

with determinant $|a_{i\alpha}| \neq 0$ *is projective.* Since the determinant is not zero, the system of equations (1) has a unique solution for the x_α and the solution has the form

$$x_\alpha = \sum_i a'_{i\alpha} x'_i, \quad \alpha = 1, 2, 3, 4,$$

where the $a'_{i\alpha}$ can be expressed in terms of the $a_{i\alpha}$ in a familiar manner. It follows that, if given points lie in a plane, i.e., satisfy a linear equation

$$\sum_{\alpha} b_{\alpha} x_{\alpha} = 0,$$

then their images satisfy the linear equation

$$\sum_{\alpha,\, i} b_{\alpha} a'_{i\alpha} x'_i = 0,$$

and, therefore, also lie in a plane. This proves the italicized assertion. Conversely, *every projective transformation is linear.* The proof of this property of projective transformations is our next goal.

Let x, y, u, v be four distinct points on a line. We say that the pair of points x, y *divides* the pair of points u, v if the points u, v belong to different intervals with endpoints x, y. Simple calculations show that *the property of division of pairs is reciprocal,* i.e., if the pair x, y divides the pair u, v, then the pair u, v also divides the pair x, y.

The pair of points $x + \lambda y$, $x - \lambda y$ (for $\lambda \neq 0$) divides the pair x, y and this division of pairs is said to be *harmonic.* Harmonic division of pairs is also reciprocal, i.e., if the pair u, v divides the pair x, y harmonically, then the pair x, y also divides the pair u, v harmonically. Harmonic division of pairs has a simple geometric interpretation. Suppose that the pair u, v divides the pair x, y harmonically. Through the point u we draw a line, different from the line xy, and on this line we choose two points a and b, different from u (see Fig. 1). The lines

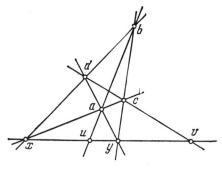

FIG. 1

xa and yb intersect in a point d. It turns out that the line cd intersects the line xy in the point v, as we now proceed to show. Suppose that the line cd intersects the line xy in the point $v' = x + \mu y$. Let A be the linear transformation under which the points a and b remain fixed, while the points c and d exchange places. Such a transformation exists, since the points a, b, c, d are noncollinear. The points u and v' remain fixed under the transformation A, but the points x and y change places. Therefore

$$Au = \xi u, \quad Av' = \xi' v', \quad Ax = \eta y, \quad Ay = \eta' x$$

where ξ, ξ', η, η' are certain numbers, all different from zero. Substituting $u = x + \lambda y$ and $v' = x + \mu y$ into the first two equations, we obtain

$$\eta y + \lambda \eta' x = \xi(x + \lambda y), \quad \eta y + \mu \eta' x = \xi'(x + \mu y).$$

Consequently

$$\eta = \xi \lambda, \quad \xi = \lambda \eta',$$
$$\eta = \xi' \mu, \quad \xi' = \mu \eta'.$$

and, therefore

$$\eta = \lambda^2 \eta', \quad \eta = \mu^2 \eta'.$$

We see that $|\lambda| = |\mu|$ and, since the points u and v' are certainly distinct, we have $\lambda = -\mu$, i.e., the point v' coincides with $v = x - \lambda y$, as asserted. It follows from this geometric interpretation of harmonic division that *harmonic division of pairs is invariant under projective transformations.*

Given two pairs of points x, y and x', y' on a line, we now find a condition under which there exists a pair of points u, v which harmonically divides both the pair x, y and the pair x', y'. We must have

$$x' = u + \lambda v, \quad y' = u - \lambda v$$

since the pair u, v divides the pair x', y' harmonically. We must also have

$$u = x + \mu y, \quad v = x - \mu y$$

since the pair x, y divides the pair u, v harmonically. It follows that

$$
\begin{aligned}
x' &= (1 + \lambda) x + \mu (1 - \lambda) y, \\
y' &= (1 - \lambda) x + \mu (1 + \lambda) y.
\end{aligned}
\tag{2}
$$

Since the points x, y, x', y' are all different, we must have $|\lambda| \neq 1$ and $\mu \neq 0$. From the expressions (2) for x' and y', in terms of x and y, it is clear that the points x' and y' belong to a single interval with endpoints x and y, i.e., the pairs x, y and x', y' do not divide one another. Thus, a necessary condition for the existence of a pair u, v, which harmonically divides both of the pairs x, y and x', y', is that these pairs do not divide each other.

Suppose, now, that the pairs x, y and x', y' do not divide each other. We may assume that

$$x' = x + \lambda y, \quad y' = x + \mu y, \tag{3}$$

where λ and μ have the same sign, so that $\lambda/\mu > 0$. It is easy to see that the pair u, v, given by

$$u = x' + \sqrt{\tfrac{\lambda}{\mu}} y', \quad v = x' - \sqrt{\tfrac{\lambda}{\mu}} y', \tag{4}$$

harmonically divides both the pair x', y' and the pair x, y. The first fact is obvious from (4). To prove the second fact, we substitute (3) into (4) to obtain

$$
\begin{aligned}
u &= \left(1 + \sqrt{\tfrac{\lambda}{\mu}} \right) x + \left(\lambda + \mu \sqrt{\tfrac{\lambda}{\mu}} \right) y \\
v &= \left(1 - \sqrt{\tfrac{\lambda}{\mu}} \right) x + \left(\lambda - \mu \sqrt{\tfrac{\lambda}{\mu}} \right) y
\end{aligned}
$$

which, by suitably renormalizing coordinates, can be put in the form

$$u = x + \theta y, \qquad v = x - \theta y.$$

Thus the pair u, v harmonically divides the pair x, y and our assertion is proven.

A necessary and sufficient condition for the existence of a pair u, v which harmonically divides both the pair x, y and the pair x', y' is that the pairs x, y and x', y' do not divide each other. From this we draw the important conclusion that *the property of division of pairs is invariant under projective transformations.* Therefore, *a projective transformation carries a line interval into a line interval.*

We now return to the question of the linearity of a projective transformation. In this regard, we first prove that *if three points of a line are fixed, under a projective transformation, then so are all the points of the line.* If x and y are two points of the line, the third point has the representation $\lambda x + \mu y$. By renormalizing the coordinates of x and y, we can arrange that $\lambda = 1$ and $\mu = 1$. Therefore, without loss of generality, it may be assumed that the three fixed points of the line are $x, y, x + y$. We note that if the three points x^1, x^2, x^3 of the line are fixed under a projective transformation and if the pair x^3, x^4 divides the pair x^1, x^2 harmonically, then the point x^4 is also fixed. This is because x^4 is uniquely determined by the points x^1, x^2, x^3 and its image is uniquely determined by the images of these three points.

Obviously, the pair $x, x + 2y$ is divided harmonically by the pair $x + y, y$. In fact,

$$x = (x + y) - y, \qquad x + 2y = (x + y) + y.$$

Therefore, the point $x + 2y$ is fixed under our projective transformation. Moreover, the pair $x + y, x + 3y$ is divided harmonically by the pair $x + 2y, y$ so that the point $x + 3y$ is fixed. Similarly, the points $x + 4y, x + 5y$ and, more generally, $x + ny$, for arbitrary integral n, are fixed.

The pair $x, x + y$ is divided harmonically by the pair $x + \frac{1}{2}y, y$ and, therefore, the point $x + \frac{1}{2}y$ is fixed. We use the method just described to prove that the point $x + \frac{n}{2}y$ is fixed for arbitrary integral

n. Next we consider the pairs x, $x + \frac{1}{2} y$ and $x + \frac{1}{4} y$, y and establish the fact that the point $x + \frac{1}{4} y$ is fixed and, similarly, that every point of the form $x + \frac{n}{4}$ is fixed for arbitrary integral n. We then establish that every point $x + \frac{n}{8} y$ is fixed, and so on. As a result, continuing in this way, we find that every point of the line of the form $x + \frac{n}{2^k} y$ is fixed.

We now prove that every point w, of the line xy, is fixed under our given projective transformation. It may be assumed without loss of generality that this point is different from x and y and, therefore, has a representation $w = x + \lambda y$. If the number λ has the form $\lambda = n/2^k$, then the point w is fixed, as we have just proven. Otherwise, we choose numbers λ' and λ'', of the form $n/2^k$, which are arbitrarily close to λ and such that $\lambda' < \lambda < \lambda''$. It is not hard to see that the points $u' = x + \mu y$, for which $\lambda' < \mu < \lambda''$, form an interval with endpoints $u' = x + \lambda' y$ and $v' = x + \lambda'' y$. This interval contains fixed points, namely those corresponding to the values of μ of the form $n/2^k$. Hence, this interval goes into itself under our projective transformation. In fact, the point $w = x + \lambda y$, of this interval, goes into some point $\bar{w} = x + \bar{\lambda} y$, where $\lambda' < \bar{\lambda} < \lambda''$. λ' and λ'' can be chosen arbitrarily close to λ. Consequently, $\bar{\lambda}$ must coincide with λ, since it lies between λ' and λ'', i.e., the point w is fixed. Thus, if three points of a line are fixed under a projective transformation, then all of the points of the line are fixed.

We now show that *if four points of a plane, no three of which are collinear, are fixed under a projective transformation, then all of the points of the plane are fixed*. Let x, y, z, u be the given fixed points of the plane. The lines joining these four points in pairs go into themselves under our projective transformation. Hence, the point of intersection v of the lines xy and zu is fixed. By what was just proven, all of the points of the line xy (which contains the three fixed points x, y, v) are fixed. In the same way we can establish that all of the points of the lines joining the points x, y, z, u in pairs are fixed. Now, let w be an arbitrary point of the plane which does not lie on these lines. Draw a line through w which does not pass through the points x, y, z. This line intersects the lines xy, yz, zx in three distinct points. Since these three points are fixed, all of the points of the line and, in particular, the point w, are fixed. Similarly, it can be shown that *if five points of space, no four of which are coplanar, are fixed under a projective transformation, then all of the points of space are fixed.*

Now, let A be an arbitrary projective transformation of space. Take five points u^i, no four of which are coplanar. Denote the preimages of the points u^i by v^i, so that $u^i = Av^i$. Then no four of the v^i are coplanar. Next, let B be the linear transformation which maps the points u^i to the points v^i, so that $v^i = Bu^i$. The successive application of the transformations A and B gives the composite transformation $B \cdot A$, which is a projective transformation. This projective transformation leaves the points v^i fixed and is, therefore, the identity transformation, by what has just been proven. It follows that the transformation A is the inverse of the transformation B. But B is a linear transformation, so its inverse is also a linear transformation. We have proven that *every projective transformation is linear.*[6]

§3. DESARGUESIAN METRIZATIONS OF A PROJECTIVE SPACE

Let $\rho(x, y)$ be a *metric* defined on a subset G of a projective space P. This means that to every ordered pair of points x, y in G there is associated a nonnegative real number $\rho(x, y)$, the *distance between the points x and y,* satisfying the following conditions:

1) $\rho(x, y) = \rho(y, x)$ for all points x, y in G
2) $\rho(x, y) = 0$ if, and only if, the points x and y coincide
3) $\rho(x, y) \leqslant \rho(x, z) + \rho(y, z)$ for any three points x, y, z in G.
(the *triangle inequality*)

The distance function $\rho(x, y)$ can be used to define the length of curves. If a curve γ is given by the equation $x = x(t)$ $(a \leqslant t \leqslant b)$, we set

$$L(\gamma) = \sup_{k} \sum \rho(x(t_{k-1}), x(t_k)),$$

where $a = t_1 < t_2 < ... < t_n = b$ and the supremum is taken over all such finite partitions of the interval $[a, b]$ by the points t_k. The number $L(\gamma)$ is the *length* of γ.[7]

We say that a curve γ is a *(metric) segment* if its length equals the distance between its endpoints. A curve γ is a *geodesic* if every point of γ has a neighborhood (on γ) which is a segment. A metric $\rho(x, y)$ is called *Desarguesian* if it is defined on a domain G of a projective space and the geodesics in the metric coincide with the projective lines (more precisely, the intersections of the projective lines with G). We shall construct a large class of Desarguesian metrics in projective space, namely the σ-metrics which were introduced by Herbert Busemann. (see [7]).

The construction of Desarguesian metrics, which will be given below, involves the use of *completely additive set functions* defined on the sets of planes of the projective space. Completely additive set functions presuppose a definite topology on the space, on whose subsets the functions are defined. We shall now introduce a topology on the set of planes of projective space. A topology can be specified with the help of neighborhoods of points of the space, provided that the neighborhoods satisfy certain conditions. After defining neighborhoods, the concepts of open and closed sets are introduced in the familiar way. A set G is said to be *open* if each point of G has a neighborhood contained in G, while a set is said to be *closed* if it is the complement of an open set. The empty set and the entire space are regarded as being both open and closed. The other concepts of a topological space, such as the concept of the convergence of points, and so on, are defined in terms of open and closed sets.

Thus, to specify a topology on the space, it is sufficient to define the concept of a *neighborhood of an arbitrary point* of the space. This concept must satisfy the following conditions, already referred to above: 1) if $N(x)$ is a neighborhood of the point x and y is a point of the space in this neighborhood, then there exists a neighborhood $N(y)$ of the point y which is contained in $N(x)$, 2) If $N_1(x)$ and $N_2(x)$ are two neighborhoods of the point x, then there exists a third neighborhood $N_3(x)$ of the point x which is contained in both $N_1(x)$ and $N_2(x)$. These conditions will be satisfied if we define as a neighborhood of an arbitrary plane, with equation $a \cdot x = 0$, the set of all planes with equations $a' \cdot x = 0$, where $|a - a'| < \varepsilon$. To be more precise, suppose that a plane α has the equation

$$a \cdot x = a_1 x_1 + a_2 x_2 + a_3 x_3 + a_4 x_4 = 0.$$

Then, for each $\varepsilon > 0$, the set of all planes with equations

$$a' \cdot x = a_1' x_1 + a_2' x_2 + a_3' x_3 + a_4' x_4 = 0$$

such that

$$|a_1 - a_1'| < \varepsilon, \quad |a_2 - a_2'| < \varepsilon, \quad |a_3 - a_3'| < \varepsilon,$$
$$|a_4 - a_4'| < \varepsilon$$

is a neighborhood of α.

A function σ, defined on the sets of a topological space, is said to be *completely additive*, if

$$\sigma \left(\sum_k H_k \right) = \sum_k \sigma (H_k)$$

for every countable system of pairwise disjoint Borel sets H_k.[8]

Let X be an arbitrary set of points of a projective space P. We will denote by πX the set of planes of P which intersect the set of points X. Suppose, now, that σ is a nonnegative completely additive set function, defined on the sets of planes of the projective space P. σ will be required to satisfy the following conditions:

1) $\sigma (\pi X) = 0$ if the set X consists of a single point
2) $\sigma (\pi X) > 0$ if X contains an interval of a projective line
3) the value of the set function σ on the set of all of the planes of P is finite.

We shall now use such a set function σ to define a corresponding Desarguesian metric on the projective space P.

First of all, we define the length of an interval of a projective line. The *length of a projective line interval* $\overline{xy}$, with endpoints x and y, is defined to be the number $|xy| = \sigma(\pi xy)$. In other words, *the length of an interval is the value of the set function σ on the set of all planes intersecting the interval.* As defined in this way, the length of an interval has the usual properties, namely, it is positive, finite, and additive. The positivity and finiteness follow from conditions 2) and 3), while the additivity follows from the additivity of the set function σ and condition 1). Note that *the value of the set function σ on the set of all of the planes of P is the length of a projective line.* Thus, *all lines have one and the same length.*

We now introduce a metric by defining a distance function $\rho(x, y)$ on the pairs of points of the space P. The points x and y divide the line passing through x and y into two intervals. The length of the shorter segment $\overline{xy}$ will be called the *distance between the points x and y* and will be denoted by $\rho(x, y)$. The distance function, defined in this way, satisfies the Axioms of a Metric Space, namely, $\rho(x, y) = \rho(y, x)$, $\rho(x, y) \geqslant 0$ with equality holding if, and only if, $x = y$, and $\rho(x, y) \leqslant \rho(x, z) + \rho(y, z)$. That the first two Axioms are satisfied is an immediate consequence of our definition of $\rho(x, y)$. We next verify the third Axiom, the triangle inequality.

If two of the three points or all of the three points coincide, then the triangle inequality is obvious. Consider the case where the three points are all different and lie on the same line. Let $\overline{xy}$ denote the interval, with endpoints x and y, to which the point z belongs. Let the intervals with endpoints x, z and y, z, belonging to $\overline{xy}$, be denoted by $\overline{xz}$ and $\overline{yz}$, and let the complementary intervals be denoted by $\overline{\overline{xy}}$, $\overline{\overline{xz}}$, and $\overline{\overline{yz}}$, respectively. It is obvious that $\overline{\overline{xy}} \subset \overline{\overline{xz}}$ and $\overline{\overline{xy}} \subset \overline{\overline{yz}}$. If $\rho(x, z) = |xz|$ and $\rho(y, z) = |yz|$, then $\rho(x, y) \leqslant |xz| + |yz| = \rho(x, z) + \rho(y, z)$. If $\rho(x, z) = |\overline{\overline{xz}}|$ then $\rho(x, y) < |xz| = \rho(x, z)$ and, hence, $\rho(x, y) \leqslant \rho(x, z) + \rho(y, z)$. Similarly, if $\rho(y, z) = |\overline{\overline{yz}}|$, then $\rho(x, y) < \rho(y, z)$ and, hence, $\rho(x, y) \leqslant \rho(x, z) + \rho(y, z)$.

In the case where the three points x, y, z are not collinear, the triangle inequality follows from the Axiom of Pasch and the additivity of the set function σ. In fact, let $\overline{xyz}$ be a projective triangle and let its side $\overline{xz}$, $\overline{yz}$ be the shorter of the intervals of the corresponding lines xz and yz. Such a triangle exists. To see this, we draw a plane intersecting the complementary intervals $\overline{\overline{xz}}$ and $\overline{\overline{yz}}$. This plane intersects one of the intervals joining the points x, y and we choose the complementary interval as the third side of the triangle $\overline{xyz}$. According to the Axiom of Pasch, a plane α intersecting the side $\overline{xy}$ of the triangle $\overline{xyz}$, intersects at least one of the two sides $\overline{xz}$, $\overline{yz}$. Consequently, $\pi\,xy \subset \pi\,xz \cup \pi\,yz$. Therefore, by the additivity of the set function σ, we have $|\overline{xy}| \leqslant |\overline{xz}| + |\overline{yz}|$. However, $\rho\,(x,\,z) = |\overline{xz}|$ and $\rho\,(y,\,z) = |\overline{yz}|$, while $\rho\,(x,\,y) \leqslant |\overline{xy}|$. It follows that $\rho\,(x,\,y) \leqslant \rho\,(x,\,z) + \rho\,(y,\,z)$. Q.E.D. A metric, defined in this way with the help of a completely additive set function σ, will be called a σ-metric.

Theorem. *Every σ-metric defined in a projective space is Desarguesian; that is, the lines of the space are the geodesics.*

PROOF. Let g be any line and let x be any point of g. We construct an infinite sequence of intervals δ_0, δ_1, δ_2, ... of the line g, containing the point x and satisfying the following conditions: 1) Every interval δ_k is contained in the preceding interval δ_{k-1} and 2) The intersection of all of the intervals consists of the single point x. Let l_0 be the length of the line g. By the complete additivity of the set function σ, we have

$$\sigma(\pi\delta_0) = \sum_k \delta\pi(\delta_{k-1} \smallsetminus \delta_k),$$

which implies that

$$\sigma\,(\pi\delta_{k_0}) = \sum_{k > k_0} \sigma\pi\,(\delta_{k-1} \smallsetminus \delta_k) < l_0/2$$

for sufficiently large k_0. This means that the length of the interval δ_{k_0} is less than the length of the complementary interval and, hence, equals the distance $\rho\,(x,\,y)$ between the endpoints of the interval. Thus, x has a neighborhood, on g, which is a segment. Since the choice of the point x was arbitrary, the line g is a geodesic. The σ-metric is Desarguesian and the Theorem is proven.

We next consider a σ-metric defined on an arbitrary convex domain G (i.e., convex open subset) of a projective space P. In this case, the function σ is defined on the set of planes intersecting G. We retain the first two conditions imposed on the set function σ, but we replace the third condition (finiteness on the set of all planes of the space P) by the following requirement: $3'$) σ $(\pi H) < \infty$ for every set H contained in G, whose closure does not intersect the boundary of G.

Just as in the definition of a σ-metric on the whole space P, we begin by defining the lengths of intervals contained in G. We then define the distance between points of the domain G as the length of the interval joining these points in G, and then prove that the distance $\rho(x, y)$ defined in this way, satisfies the Axioms of a metric space and is a Desarguesian metric. It should be noted that the uniqueness of the interval joining two points x and y only simplifies the corresponding treatment.

§4. REGULAR DESARGUESIAN METRICS IN THE TWO-DIMENSIONAL CASE

Our basic tool for investigating general Desarguesian metrics will be to approximate them in a special way by regular Desarguesian metrics. Accordingly, in this section, we consider regular Desarguesian metrics in an arbitrary convex domain G of projective space. For convenience, we introduce nonhomogeneous coordinates x, y, z in the domain G, in the following way. Let α be the plane given by the equation $a \cdot x = 0$ in homogeneous coordinates. For every point of space which does not lie in the plane α, we put

$$x = \frac{x_1}{a \cdot x}, \quad y = \frac{x_2}{a \cdot x}, \quad z = \frac{x_3}{a \cdot x}$$

The numbers x, y, z are uniquely determined by the chosen point and are called its *nonhomogeneous coordinates*. It is not hard to see that planes are given by nonhomogeneous linear equations in non-homogeneous coordinates and lines by a system of two independent such equations. By assigning to each point with nonhomogeneous coordinates x, y, z the point of euclidean space with the same coordinates, we obtain a mapping of projective space, with the plane α omitted, onto euclidean space. This mapping carries planes into planes and lines into lines. Convex domains of projective space that do not intersect the plane α go into convex domains of euclidean space. Moreover, a Desarguesian metric on projective space induces a Desarguesian metric on euclidean space. Consequently, the local investigation of Desarguesian metrics in projective space reduces to the consideration of such metrics in euclidean space. For expository convenience, we shall denote the cartesian coordinates of a point in euclidean space by x_1, x_2, x_3.

A metric, in euclidean space with cartesian coordinates x_i, will be called a *regular* (*Finsler*) *metric* if it is specified by means of a line element

$$ds = F\,(x, dx)$$

where F is a smooth function of the variables x, dx (for $dx \neq 0$) which is positive homogeneous of degree 1, convex, and even in the variable dx. This means that 1) If $y \neq 0$, then $F(x, y) > 0$ for every point x in the domain of definition of the metric, 2) $F\,(x, \theta y) = |\theta|\,F\,(x, y)$ for every number $y \neq 0$, and 3) If $y \neq 0$, then $d_y^2\,F\,(x, y) > 0.$[9]

To specify a metric with the help of a line element we proceed as follows. First, we define the length $L(\gamma)$ of a smooth curve γ by means of the customary formula

$$L(\gamma) = \int_\gamma F\,(x, dx).$$

Second, we define the *distance* $\rho\,(x, y)$ of two points x and y by

$$\rho\,(x, y) = \sup L(\gamma)\,,$$

where the supremum is taken over all smooth curves joining x to y.

Theorem: *In the two-dimensional case, every regular Desarguesian metric is a σ-metric and its line element has the representation*

$$ds = \int_\omega \gamma\,(g)\,|\xi\,dx|\,d\xi.$$

Here, g is the line passing through the point x perpendicular to the unit vector ξ, $\gamma\,(g)$ is a nonnegative function, and the integration is taken over the unit circle ω: $|\xi| = 1$. The set function σ specifying this metric is defined by the formula $d\sigma = \gamma(g)\,dp\,d\xi$, where p and ξ are the coefficients in the equation $\xi \cdot x - p = 0$ of the line g.[10]

PROOF. Let the given regular Desarguesian metric be given by $ds = F(x, dx)$. Since the metric is Desarguesian, the geodesics are straight lines. The geodesic joining the points a and b is an extremal of the functional

$$s = \int_{(a)}^{(b)} F\,(x, \dot{x})\,dt,$$

and, therefore, satisfies the Euler Equations

$$\frac{\partial F}{\partial x_k} - \frac{d}{dt}\frac{\partial F}{\partial \dot{x}_k} = 0 \quad (k = 1, 2),$$

or, carrying out the indicated differentiation

$$\frac{\partial F}{\partial x_k} - \sum_\alpha \dot{x}_\alpha \frac{\partial^2 F}{\partial \dot{x}_k\,\partial x_\alpha} - \sum_\alpha \ddot{x}_\alpha \frac{\partial^2 F}{\partial \dot{x}_k\,\partial \dot{x}_\alpha} = 0,\; k = 1, 2,$$

where, in each equation, the summation is extended over $\alpha = 1, 2$.[11] Suppose that we choose the parameter t to be the euclidean arclength of the extremal. Then, since the extremal is a straight line and is, therefore, specified by equations which are linear in t, the terms involving $\ddot{x}_\alpha$ vanish and we arrive at the following system of equations for the function F:

$$\frac{\partial F}{\partial x_k} - \sum_\alpha \dot{x}_\alpha \frac{\partial^2 F}{\partial \dot{x}_k \partial x_\alpha} = 0, \ k = 1, 2. \tag{1}$$

Differentiating (1) with respect to $\dot{x}_k$, we obtain

$$\sum_\alpha \dot{x}_\alpha \frac{\partial}{\partial x_\alpha} \left(\frac{\partial^2 F}{\partial \dot{x}_k^2} \right) = 0, \ k = 1, 2. \tag{2}$$

Now, let g be an arbitrary line and take it to be the x_1-axis. Then, along this line, that is, where $x_2 = 0$ and therefore $\dot{x}_2 = 0$, the second equation of (2) becomes

$$\frac{\partial}{\partial x_1} \left(\frac{\partial^2 F}{\partial \dot{x}_k^2} \right) = 0. \tag{3}$$

It follows from (3) that $\partial^2 F / \partial \dot{x}_2^2$ is constant along the line g and, consequently, is a function of the line g. Let this function be denoted by $\gamma \, (g)$.

Consider the metric specified by the line element

$$ds = \int_\omega \gamma \, (g) \, |\xi \, dx| \, d\xi,$$

where g is the line passing through the point x perpendicular to the unit vector ξ, and the integration is extended over the unit circle ω: $\| \xi \| = 1$. The metric specified by this line element is a σ-metric. In fact, the expression

$$\int_\omega \gamma \, (g) \, |\xi \, dx| \, d\xi$$

represents the value of the function σ, defined by the formula

$$\sigma = \int \gamma \, dp \, d\xi,$$

on the set of lines intersecting the element dx (see Note (10) for the calculation).

Let

$$F_1(x, \dot{x}) = \int_\omega \gamma \, |\xi \dot{x}| \, d\xi.$$

To prove the Theorem, we need only show that $F = cF_1$, where c is a constant. To this end, we take a small isosceles triangle uvw, with base uv of euclidean length δ on the line g, and small base angles, equal to α. Let Δ be the difference between the sum of the lengths of the lateral sides of the triangle and the length of its base in the σ-metric specified by the line elements $ds = F_1(x, \dot{x})$. This difference Δ can be expressed in two ways, in terms of the function F_1 specifying the line element and also in terms of the function γ.

We first express Δ in terms of γ. The quantity Δ is equal to the value of the set function σ on the set of lines intersecting the lateral sides of the tirangle. Hence

$$\Delta = \int \gamma \, dp \, d\xi,$$

where the integration is carried out over the set of lines g intersecting both lateral sides $\overline{uw}$ and $\overline{vw}$ of the triangle. Because γ is continuous and α and δ are small, we have the approximation

$$\Delta \simeq \gamma(g) \int dp \, d\xi.$$

For sufficiently small α and δ the integral on the right is proportional to the altitude of the triangle $\simeq \alpha\delta/2$ (the integration with respect to p) and to the base angle α (the integration with respect to ξ). We therefore have the approximation

$$\Delta \simeq c_1 \gamma(g) \, \delta \alpha^2$$

for sufficiently small α and δ. This approximation gives the precise value if we take the limit first as $\alpha \to 0$ and then as $\delta \to 0$.

We now express the difference Δ in terms of the function F_1 specifying the line element. Taking x_1 as the independent variable, we have

$$ds = F_1\,(x_1, x_2, \dot{x}_1, \dot{x}_2)\,dt = f\,(x_1, x_2, \dot{x}_2)\,dx_1.$$

Using the Taylor Formula, we next expand the function f about the point $x_1 = x_2 = \dot{x}_2 = 0$ to obtain

$$f(x_1, x_2, \dot{x}_2) = f + \frac{\partial f}{\partial x_1}\,x_1 + \frac{\partial f}{\partial x_2}\,x_2 + \frac{\partial f}{\partial \dot{x}_2}\,\dot{x}_2 +$$

$$\frac{1}{2}\left(\frac{\partial^2 f}{\partial x_1^2}\,x_1^2 + \dots + \frac{\partial^2 f}{\partial \dot{x}_2^2}\,\dot{x}_2^2 \right) + (*).$$

In the right-hand side of the formula, the function f and its derivatives are evaluated at the foot of the altitude of the triangle for $\dot{x}_2 = 0$, and $(*)$ denotes terms of higher order. We have

$$\Delta = \int_{-\delta/2}^{0} f\,(x_1, x_2, \tan\,\alpha)\,dx_1 + \int_{0}^{\delta/2} f\,(x_1, x_2, -\tan\,\alpha)\,dx_1 -$$

$$\int_{-\delta/2}^{\delta/2} f\,(x_1, 0, 0)\,dx_1.$$

By substituting the expression for f into this formula, we see that the main contribution to Δ is made, for small δ/α, by the term $\frac{\partial^2 f}{\partial \dot{x}_2^2}\,\dot{x}_2$, as the other terms are all of higher degree in δ. Therefore, for fixed α and sufficiently small δ, we have the approximation

$$\Delta \simeq c_2 \frac{\partial^2 f_1}{\partial \dot{x}_2^2}\,\alpha^2 \delta.$$

Comparing the two expressions for Δ just obtained, we see that

$$\gamma(g) \simeq c\,\frac{\partial^2 f}{\partial \dot{x}_2^2}\,.$$

Consequently, taking the limit, first as $\delta \to 0$ and then as $\alpha \to 0$, we obtain

$$\gamma(g) = c\,\frac{\partial^2 F_1}{\partial \dot{x}_2^2}\,.$$

or, recalling that $\gamma(g)$ was defined by $\gamma(g) = \partial^2 F/\partial \dot{x}_2^2$, we finally have

$$\frac{\partial^2 F}{\partial \dot{x}_2^2} = c\,\frac{\partial^2 F_1}{\partial \dot{x}_2^2}\,. \tag{4}$$

Since both the line g and the point of g to which the triangle shrinks in passing to the limit were arbitrary, it follows from (4) that

$$d^2 F = c\,d^2 F_1\,.$$

But this means that the functions F and cF_1 differ only by a homogeneous linear term in $\dot{x}_1$ and $\dot{x}_2$. Because of the symmetry of both functions in $\dot{x}$, this term must equal zero. Therefore

$$F = cF_1,$$

that is, the line element of the given Desarguesian metric is of the form

$$ds = c \int_\omega \gamma\,|\xi\,dx|\,d\xi,$$

and is, therefore, a σ-metric.

§5. AVERAGING DESARGUESIAN METRICS

Let $\rho_1(x, y)$ and $\rho_2(x, y)$ be Desarguesian metrics given in a domain G, and let λ, μ be nonnegative numbers. Then the distance function $\rho(x, y)$ in the domain G, defined by the formula

$$\rho(x, y) = \lambda \rho_1(x, y) + \mu \rho_2(x, y), \quad \lambda, \mu \geqslant 0.$$

obviously satisfies the axioms for a metric space and, therefore, specifies a metric in the domain G. We wish to see that *the metric $\rho(x, y)$ is again a Desarguesian metric*. To show this, let g be an arbitrary line and x be an arbitrary point of g. We need only prove that there exists an interval δ of the line g, containing x as an interior point, which is a metric segment in the metric $\rho(x, y)$, i.e., such that the length of δ equals the distance between its endpoints.

Since the metric ρ_1 is Desarguesian, there exists an interval δ_1 on the line g, containing the point x in its interior, which is a segment in the metric ρ_1. Similarly, there exists a segment δ_2, again containing the point x in its interior, which is a segment in the metric ρ_2. The common part δ of the intervals δ_1 and δ_2 is again an interval, containing x as an interior point, which is a metric segment in both the metrics ρ_1 and ρ_2.[12] Let x' and x'' be the endpoints of the interval δ. We now calculate the length of the interval δ in the metric ρ. By definition, this is

$$L(\delta) = \sup \sum_k \rho(x_{k-1}, x_k),$$

where the x_k are successive points of the interval δ. Moreover,

$$\sum_k \rho(x_{k-1}, x_k) = \lambda \sum_k \rho_1(x_{k-1}, x_k) + \mu \sum_k \rho_2(x_{k-1}, x_k).$$

But the interval δ is a metric segment in both the metrics ρ_1 and ρ_2 so that

$$\sum_k \rho_1(x_{k-1}, x_k) = \rho_1(x', x''), \quad \sum_k \rho_2(x_{k-1}, x_k) = \rho_2(x', x'').$$

Consequently

$$L(\delta) = \lambda \rho_1(x', x'') + \mu \rho_2(x', x'') = \rho(x', x''),$$

that is, the interval δ is a segment in the metric $\rho(x, y)$. Q.E.D.

Next, let $\rho(x, y)$ be an only continuous Desarguesian metric defined in a convex domain G' of euclidean space. Suppose that every straight line interval, contained in G', is a segment in the metric $\rho(x, y)$. Suppose, further, that the length of an interval lying in G'. in this metric $\rho(x, y)$, is bounded by a constant D which is independent of the particular interval chosen. Let G be a convex domain, lying strictly inside G', and let δ_0 be the distance from G to the boundary of G'. In the domain G we define a distance function $\rho_\delta(x, y)$ by the formula

$$\rho_\delta(x, y) = \lambda \int_{\|r\| < \delta} \varphi(r) \rho(x + r, y + r)\, dr, \tag{1}$$

where x and y are points of G and the integration is extended over the solid ball $\|r\| < \delta \leqslant \delta_0/3$. The function $\varphi(r)$ is defined by the formula

$$\varphi(r) = \begin{cases} \exp(\|r\|^2 - \delta^2)^{-1} & \text{if } \|r\| < \delta \\ 0 & \text{if } \|r\| \geqslant \delta, \end{cases}$$

and the normalization factor λ is determined by the condition

$$\lambda \int_{\|r\| < \delta} \varphi(r)\, dr = 1. \tag{2}$$

Theorem. *The function $\rho_\delta(x, y)$ defines a Desarguesian metric in the domain G, which converges uniformly to the metric $\rho(x, y)$, as $\delta \to 0$. The function*

$$F_\delta(x, \xi) = \lim_{h \to 0} \frac{1}{h} \rho_\delta(x, x + \xi h)$$

exists for arbitrary x, ξ and is uniformly bounded in G. Moreover, $F_\delta(x, \xi)$ is continuous in (x, ξ) and is positive homogeneous, symmetric and convex in ξ. The metric specified by the line element

$$ds = F_\delta(x, dx)$$

coincides with the metric $\rho_\delta(x, y)$.

PROOF. The fact that the distance function $\rho_\delta(x, y)$ defines a Desarguesian metric in the domain G is established in the same way as for the metric $\lambda\rho_1 + \mu\rho_2$, in the preceding proof. Moreover, $\rho_\delta(x, y) \rightarrow \rho(x, y)$, uniformly in x, y, as $\delta \rightarrow 0$, since the function $\rho(x, y)$ is continuous in the domain G' and, hence, is uniformly continuous in a $2\delta_0/3$-neighborhood of the domain G. We now prove the remaining assertions of the Theorem in more detail.

First of all, we note that the function $\rho_\delta(x + z, y + z)$ is continuously differentiable in z. In fact

$$\partial \rho_\delta(x + z, y + z)/\partial z_k$$

$$= \lambda \int_{\|r\| < \delta} \frac{\partial \varphi(r)}{\partial r_k} \rho(x + z + r, y + z + r) dr$$

since $\varphi(r) = 0$ if $\|r\| \geqslant \delta$.[13]

Next, consider the function $\rho_\delta(x, x + a \cdot \xi)$ of the variables x and a, for a fixed vector ξ, $\|\xi\| = 1$, where x belongs to the domain G and a satisfies $\delta_0/3 \leqslant a \leqslant 2\delta_0/3$. In a similar way (see Note (13)), the function $\rho_\delta(x, x + a \cdot \xi)$ is continuously differentiable in x.

For fixed x and ξ the function $\rho_\delta(x, x + a \cdot \xi)$ is monotonically increasing in a, since it represents the length of the segment from x to $x + a \cdot \xi$ in the metric $\rho_\delta(x, y)$. A monotone function has a derivative almost everywhere. Moreover, a point a' can be found for which this derivative satisfies the inequality

$$\partial \rho_\delta(x, x + a \cdot \xi)/\partial a \big|_{a=a'}$$

$$\leqslant \frac{3}{\delta_0} \left(\rho_\delta\left(x, x + \frac{2\delta_0 \cdot \xi}{3}\right) - \rho_\delta\left(x, x + \frac{\delta_0 \cdot \xi}{3}\right) \right).$$

Since the length of a segment is additive and $\rho_\delta\,(x,\,y)$ is the length of the segment from x to y in the metric ρ_δ, we have

$$\rho_\delta(x,\,x + \xi\cdot h) = \rho_\delta(x,\,x + a'\cdot\xi) - \rho_\delta(x + \xi\cdot h,\,x + a'\cdot\xi)$$

and, therefore,

$$\frac{1}{h}\,\rho_\delta(x,\,x + \xi\cdot h)$$

$$= \frac{1}{h}\,(\rho_\delta\,(x,\,x + a'\cdot\xi) - \rho_\delta\,(x,\,x + (a' - h)\cdot\xi))$$

$$- \frac{1}{h}\,(\rho_\delta\,(x + h\cdot\xi,\,x + \xi\cdot h + (a' - h)\cdot\xi)$$

$$- \rho_\delta\,(x,\,x + (a' - h)\cdot\xi)).$$

As $h \to 0$, the first term on the right-hand side of the equation approaches a definite limit, since the function $\rho_\delta(x,\,x + a\cdot\xi)$ has a derivative with respect to a at $a = a'$. The second term also converges to a limit, since the function $\rho_\delta\,(x,\,x + a\cdot\xi)$ is continuously differentiable in x. Hence, the limit of the left-hand side of the equation exists, as $h \to 0$, and satisfies the formula

$$F_\delta(x,\,\xi) = \partial\,\rho_\delta(x,\,x + a\xi)/\partial a\,|_{a=a'}$$

$$- \sum_k \xi_k\,\partial\,\rho_\delta(x,\,x + a\xi)/\partial x_k\,|_{a=a'}. \qquad (3)$$

From this formula it is clear that the function $F_\delta(x,\,\xi)$ has a uniform bound in G.

We now show that the derivative $\partial\,\rho_\delta(x,\,x + a\xi)$ exists for every a such that $\delta_0/3 \leqslant a \leqslant 2\delta_0/3$, and is continuous in a for fixed x and ξ. We have

$$\frac{1}{h}\,(\rho_\delta(x,\,x + (a + h)\xi) - \rho_\delta(x,\,x + a\xi))$$

$$= \frac{1}{h}\,\rho_\delta\,(x,\,x + h\xi)$$

$$+ \frac{1}{h}\,(\rho_\delta\,(x + h\xi,\,x + (a + h)\,\xi) - \rho_\delta\,(x,\,x + a\xi)).$$

As $h \to 0$, the first term on the right approaches a definite limit, which does not depend on a. The second term approaches the limit

$$\sum_h \xi_h \, \partial \, \rho_\delta(x, \, x + a\xi)/\partial x_h$$

which is continuous in a. Consequently, as asserted, the limit of the left-hand side of the equation, namely $\partial \rho_\delta(x, \, x + a\xi), /\partial a$, exists and is continuous in a.

We next show that the function

$$F_\delta(x, \xi) = \lim_{h \to +0} \frac{1}{h} \, \rho_\delta(x, \, x + \xi h)$$

is continuous in x, ξ. Fixing a small $\varepsilon > 0$, we choose numbers a_1 and a_2, close to a, such that the change in the derivative $\partial \, \rho_\delta(x, \, x + a \cdot \xi)/\partial a$ on the interval (a_1, a_2) is less than ε. As $x' \to x$ and $\xi' \to \xi$,

$$\rho_\delta(x', \, x' + a_1'\xi) \to \rho_\delta(x, \, x + a_1\xi),$$
$$\rho_\delta(x', \, x' + a_2'\xi) \to \rho_\delta(x, \, x + a_2\xi),$$

which implies that, if x' is sufficiently close to x and ξ' is sufficiently close to ξ, then a point a' can be found for which the derivative $\partial \, \rho_\delta(x', \, x' + a \cdot \xi)/\partial a$ at $a = a'$ differs from the derivative $\partial \, \rho_\delta(x, \, x + a \cdot \xi)/\partial a$ by less than a. Moreover, it can also be assumed that

$$|\partial \, \rho_\delta(x', \, x' + a\xi')/\partial x' - \partial \, \rho_\delta(x, \, x + a\xi)/\partial x| < 2\varepsilon.$$

We conclude from equation (3), which is valid for arbitrary x and ξ, that

$$|F_\delta(x', \xi') - F_\delta(x, \xi)| < 2\varepsilon,$$

thereby proving the continuity of F_δ in (x, ξ).

The fact that the function $F_\delta(x, \xi)$ is positive homogeneous of degree 1 in ξ follows from the very definition of the function:

$$F_\delta(x, \xi) = \lim_{h \to 0} \frac{1}{h} \rho_\delta(x, x + \xi h).$$

To verify that the function $F_\delta (x, \xi)$ is even in ξ, i.e., that $F_\delta (x, \xi) = F_\delta (x, -\xi)$, we note that the derivative of the function $\rho_\delta(x, x + h\cdot\xi)$, with respect to h at $h = 0$ is equal to $F_\delta (x, \xi)$ if the derivative is evaluated from the side $h > 0$, and is equal to $F_\delta (x, -\xi)$ if the derivative is evaluated from the side $h < 0$. But, as shown above, the function $\rho_\delta(x, x + h\cdot\xi)$ is differentiable with respect to h. Therefore, the values of the one-sided derivatives are the same and $F_\delta(x, \xi) = F_\delta(x, -\xi)$.

We now show that the metric ρ_δ is specified by the line element

$$ds = F_\delta (x, dx).$$

Let ρ_δ' denote the metric specified by the line element ds. For fixed x and ξ we have

$$\frac{d}{dt} \rho_\delta(x, x + t\xi) = F_\delta(x + t\xi, \xi)$$

and, therefore,

$$\rho_\delta(x, x + a\xi) = \int_0^a F_\delta(x + t\xi, \xi)\, dt.$$

The distance between the points x and $x + a \cdot \xi$ appears on the left-hand side of this equation, while the length of the segment joining these points, measured in the metric ρ_δ', appears on the right-hand side. Let γ be a smooth curve. Then the length of γ is the same in both the metric ρ_δ and the metric ρ_δ'. To see this, we note that, by definition, the length of the curve γ in the metric ρ_δ is given by

$$L(\gamma) = \sup \sum_k \rho_\delta(x_{k-1}, x_k).$$

It follows that there exists a sequence of partitions of the curve γ, by points x_k^n such that

$$L(\gamma) = \lim_{n \to \infty} \sum_k \rho_\delta \left(x_{k-1}^n, x_k^n \right).$$

Moreover, without loss of generality, we can assume that the Euclidean distances $\left| x_{k-1}^n, x_k^n \right|$, between consecutive points of the decomposition, are uniformly small for sufficiently large n. The length of the curve γ in the metric ρ_δ' is given by

$$L'(\gamma) = \int_\gamma F_\delta(x, dx).$$

Suppose that we inscribe a polygonal curve Γ^n with vertices x_k^n in the curve γ. Then

$$\int_{\Gamma^n} F_\delta(x, dx) \to \int_\gamma F_\delta(x, dx)$$

as $n \to \infty$, since the function $F_\delta(x, \xi)$ is continuous in (x, ξ). But the integral along a segment of Γ^n equals the distance between its endpoints in the metric ρ_δ and, therefore

$$\sum_k \rho_\delta(x_{k-1}^n, x_k^n) \to \int_\gamma F_\delta(x, dx).$$

As $n \to \infty$, which implies that $L(\gamma) = L'(\gamma)$. By definition, the distance between two points x and y in the metric ρ_δ' is the greatest lower bound of the lengths of the curves joining the given points. The distance between the points x and y in the metric ρ_δ is also the greatest lower bound of the lengths of the curves joining the segments, since the segment xy is the shortest join in the metric ρ_δ. Since the lengths of curves are equal in the metrics ρ_δ and ρ_δ', distances are also the same in these metrics and our assertion is proved.

Finally, we prove that the function $F_\delta(x, \xi)$ is convex in ξ. Let x be an arbitrary point and lay off from x two vectors $\theta\lambda\xi_1$ and $\theta\mu\xi_2$, λ, μ, $\theta > 0$. If x_1 and x_2 are the endpoints of these two vectors, then

$$\rho'_\delta(x_1, x_2) \leqslant \rho'_\delta(x, x_1) + \rho'_\delta(x, x_2).$$

We divide this inequality by θ and pass to the limit as $\theta \to 0$ with fixed x, ξ_1, ξ_2, λ, μ. Then, taking account of the continuity of the function $F_\delta(x, \xi)$, specifying the metric ρ'_δ, we obtain the formula

$$F_\delta(x, \lambda\xi_1 + \mu\xi_2) \leqslant \lambda F_\delta(x, \xi_1) + \mu F_\delta(x, \xi_2)$$

expressing the convexity of the function $F_\delta(x, \xi)$ in ξ. The proof of the theorem is now complete.

§6. REGULAR APPROXIMATION OF DESARGUESIAN METRICS

In connection with Hilbert's Fourth Problem, we will consider complete continuous Desarguesian metrics. A metric defined in a domain G of a projective space is said to be complete if convergence of a sequence of points x_n in the Cauchy sense implies convergence in the ordinary sense. This means that if $\rho(x_m, x_n) \to 0$ as m, $n \to \infty$, then there exists a point x_0 such that $\rho(x_0, x_n) \to 0$.

We will distinguish between three kinds of Desarguesian metrics. A Desarguesian metric defined in a convex domain G, whose closure fails to intersect some plane, will be called a *metric of hyperbolic type*. The metric of a Lobachevskian space is a metric of this type. It is defined in the interior of an ellipsoid in the Cayley-Klein realization of Lobachevskian geometry.

A Desarguesian metric which is defined in a domain of a projective space, which is the complement of some plane α, will be called a *metric of parabolic type.* The metric of Euclidean space is a metric of parabolic type.

Finally, a Desarguesian metric defined in all of projective space will be called a *metric of elliptic type.* The standard metric of the elliptic (Riemann) noneuclidean geometry is a metric of this type.

In the case of a metric of elliptic type, the distance between a pair of points x and y is uniformly bounded, that is, there exists a constant δ_0 such that $\rho(x, y) \leqslant \delta_0$. In the case of metrics of the parabolic and hyperbolic types, the distance $\rho(a, x) \to \infty$ as the point x gets arbitrarily close (in the sense of the topology of the space) to the boundary of the domain of definition of the metric.

Theorem. *A complete continuous Desarguesian metric, defined in a projective space, can be approximated by a regular Desarguesian metric (of class C^∞) which converges uniformly to the given metric on every compact set.*

We first prove the theorem for metrics of hyperbolic and parabolic type. In this case, it is convenient to consider the metric to be defined in Euclidean space, regarding the plane α that does not intersect the domain of definition of the metric as being the plane at infinity. Let $\rho(x, y)$ be the metric under consideration and let G_0 be its domain of definition. G_0 is a bounded convex domain for metrics of hyperbolic type and the entire space for metrics of parabolic type.

Just as was done in Section 5, we construct a metric $\rho_\delta(x, y)$ by averaging the metric $\rho(x, y)$. As $\delta \to 0$ the metric ρ_δ converges uniformly to the metric ρ on every convex domain which, together with its boundary, belongs to G_0. The metric ρ_δ is specified by the line element

$$ds = F_\delta(x, dx)$$

where F_δ is a nonnegative function which is continuous in (x, ξ) and which is even and convex in dx. Let

$$F_{\delta\varepsilon}(x, \xi) = \lambda \int\limits_{\|r\| < \varepsilon} \varphi(r) F_\delta(x + r, \xi)\, dr$$

where

$$\varphi(r) = \exp\left(\|r\|^2 - \varepsilon^2\right)^{-1} \quad \text{if} \quad \|r\| < \varepsilon$$

and the normalization factor λ is determined from the condition

$$\lambda \int_{\|r\| < \delta\varepsilon} \varphi(r)\, dr = 1.$$

The function $F_{\delta\varepsilon}$, just as is the function F_δ, is nonnegative, continuous in (x,ξ), and even and convex in ξ. Moreover, $F_{\delta\varepsilon}$ is infinitely differentiable with respect to x. This differentiability with respect to x follows from the fact that the function φ is infinitely differentiable and vanishes, together with all of its derivatives, when $\|r\| \geq \varepsilon$.

Consider the metric $\rho_{\delta\varepsilon}(x, y)$ specified by the line element

$$ds = F_{\delta\varepsilon}(x,\, dx).$$

It is obvious that the metric $\rho_{\delta\varepsilon}$ converges to the metric ρ_δ as $\varepsilon \to 0$. We now show that $\rho_{\delta\varepsilon}$ is Desarguesian. To this end, let x and y be two arbitrary points of a domain G, let d be the straight line segment joining them, and let γ be an arbitrary smooth curve joining the points x and y in G_0. Moreover, let d_r and γ_r be the segment and the curve obtained from d and γ by translation through the vector r. Since the metric ρ_δ is Desarguesian, every straight line segment in G_0 is a shortest join. Consequently

$$\int_{d_r} F_\delta(x,\, dx) \leq \int_{\gamma_r} F_\delta(x,\, dx),$$

or

$$\int_{d} F_\delta(x + r,\, dx) \leq \int_{\gamma} F_\delta(x + r,\, dx).$$

Multiplying this inequality by $\lambda \varphi(r)$ and integrating with respect to dr, we obtain

$$\int_d F_{\delta\varepsilon}(x, dx) \leqslant \int_\gamma F_{\delta\varepsilon}(x, dx),$$

that is, the segment d is the shortest join in the metric $\rho_{\delta\varepsilon}$. This proves that the metric $\rho_{\delta\varepsilon}$ is Desarguesian.

Next, let $A(\theta)$ denote the linear transformation which consists in performing consecutive rotations about the x_1, x_2, x_3 coordinate axes through small angles θ_1, θ_2, θ_3. Let

$$F_{\delta\varepsilon\alpha}(x, \xi) = \lambda \int_{\|\theta\| < \alpha} \varphi(\theta) F_{\delta\varepsilon}(Ax, A\xi) \, d\theta, \tag{1}$$

where $d\theta = d\theta_1 d\theta_2 d\theta_3$ and $\|\theta\| = (\theta_1^2 + \theta_2^2 + \theta_3^2)^{1/2}$. It is obvious that the function $F_{\delta\varepsilon\alpha}(x, \xi)$ converges to $F_{\delta\varepsilon}(x, \xi)$ as $\alpha \to 0$. We now show that the function $F_{\delta\varepsilon\alpha}$ is infinitely differentiable with respect to both variables x and ξ. We first note that it is infinitely differentiable with respect to x, since the function $F_{\delta\varepsilon}$ has this property.

To investigate the differentiability of $F_{\delta\varepsilon\alpha}$ with respect to ξ, we let $1 + B dt$ denote the orthogonal transformation which carries the vector ξ into the infinitesimally near vector $\xi + d\xi$. Then

$$F_{\delta\varepsilon\alpha}(x, \xi + d\xi) = \lambda \int_{\|\theta\| < \alpha} \varphi(\theta) F_{\delta\varepsilon}(Ax, A(1 + B \, dt)\xi) \, d\theta. \tag{2}$$

We now use the orthogonal transformation $1 + B dt$ to go over to new Cartesian coordinates x. This causes the right-hand side of the above equation to transform in the following way. The argument Ax goes into $Ax + A_1 x \, dt$, $\varphi(\theta)$ goes into $\varphi(\theta) + \varphi_1(\theta) dt$, and $A(1 + B \, dt)\xi$ goes into $A\xi$. As for the region of integration, we can ignore its change since the function φ vanishes outside the domain. Thus, our change of variables transforms the right-hand side of equation (2) into

$$\int_{\|\theta\|<\alpha} (\varphi(\theta) + \varphi_1(\theta)\,dt)\,F_{\delta\varepsilon}(Ax + A_1x\,dt,\ A\xi)\,d\theta.$$

We now form the ratio

$$\frac{1}{t}\left(F_{\delta\varepsilon\alpha}(x,\ \xi + d\xi) - F_{\delta\varepsilon\alpha}(x,\ \xi)\right)$$

which obviously approaches the limit

$$\int_{\|\theta\|<\alpha} \varphi_1(\theta)\,F_{\delta\varepsilon}(Ax,\ A\xi)\,d\theta$$

$$+ \int_{\|\theta\|<\alpha} \varphi(\theta)\,A_1x\,\nabla F_{\delta\varepsilon}(Ax,\ A\xi)\,d\theta,$$

where ∇ denotes the gradient with respect to the variables Ax. This proves the first order differentiability of $F_{\delta\varepsilon\alpha}$ with respect to ξ. Differentiability of arbitrary order is proven similarly. To prove the existence of the mixed derivatives with respect to x and ξ, we first differentiate equation (1) with respect to x and then establish differentiability with respect to ξ by the method just described.

We now show that the metric $\rho_{\delta\varepsilon\alpha}$, specified by the line element

$$ds = F_{\delta\varepsilon\alpha}\ (x,\ dx),$$

is Desarguesian. Let d be an arbitrary straight line segment and γ an arbitrary smooth curve joining its endpoints. Let d_A and γ_A be the segment and curve obtained by applying the transformation A to the segment d and the curve γ. Since the metric $\rho_{\delta\varepsilon}$ is Desarguesian, the segment d_A is the shortest join in $\rho_{\delta\varepsilon}$ and, therefore

$$\int\limits_{d_A} F_{\delta\varepsilon}(x, dx) \leqslant \int\limits_{\gamma_A} F_{\delta\varepsilon}(x, dx)$$

or

$$\int\limits_{d} F_{\delta\varepsilon}(Ax, A\, dx) \leqslant \int\limits_{\gamma} F_{\delta\varepsilon}(Ax, A\, dx).$$

Multiplying this inequality by $\lambda\varphi(\theta)$ and integrating over the region $\|\theta\| < \alpha$, we obtain

$$\int\limits_{d} F_{\delta\varepsilon\alpha}(x, dx) \leqslant \int\limits_{\gamma} F_{\delta\varepsilon\alpha}(x, dx),$$

i.e., the segment d is a shortest join in the metric $\rho_{\delta\varepsilon\alpha}$. Hence $\rho_{\delta\varepsilon\alpha}$ is a Desarguesian metric. This proves the theorem in the case of metrics of the hyperbolic and parabolic types.

We now consider a Desarguesian metric of elliptic type. In order to apply the technique of regularization of the metric, developed in Section 5, it is necessary that the shortest join of any two points in the given convex domain G belong to the domain. In the case of metrics of the hyperbolic and parabolic types, this condition holds for any convex domain G. However, in the case of a metric of elliptic type, it may happen that the straight line segment of shortest length joining two points of the domain G fails to lie entirely inside G. To avoid this difficulty, we proceed as follows. The distance between any two points in a Desarguesian metric is the greatest lower bound of the lengths of the curves joining the points. The length of a curve is defined in terms of the distances between sufficiently close points. Consequently the metric is determined as soon as it is specified in a small neighborhood of every point. Furthermore, in order that the metric ρ_δ converge to ρ, it is sufficient that it converge in a neighborhood of every point.

We now show that, in the case of a metric of elliptic type, every point has a neighborhood which is geodesically convex; that is, which contains with any two points the shortest join of the two points. In fact, let x be the given point and choose a plane that does not pass through the point x. We take a convex neighborhood N_ε of the point x, of diameter less than ε in the metric ρ. Since the metric ρ is continuous such a neighborhood can be constructed without difficulty. We must verify that, if ε is sufficiently small, then the shortest join of any two points of N_ε lies entirely inside N_ε. Suppose this assertion is false. Then there exist a sequence of numbers $\varepsilon_n \to 0$ and a sequence of domains N_{ε_n} such that a pair of points x_n and y_n can be found in the domain N_{ε_n} whose shortest join does not lie entirely inside G. This shortest join is the straight line segment $x_n y_n$, which intersects the plane σ in some point z_n. The distance between the points x_n and z_n is less than ε_n. Since the projective space is compact, we can assume without loss of generality that the sequences of points x_n and z_n converge to x_0 and z, where z belongs to σ. Since the metric ρ is continuous, and $\rho(x_n, z_n) \to 0$ as $n \to \infty$, we have $\rho(x, z) = 0$. But this is a contradiction, since the point x does not lie in the plane σ, and hence is certainly different from z. Thus every point x has a convex neighborhood in which any two points are joined by the shortest segment in the metric ρ.

To define the metric ρ_δ in Section 5, we began (in Sec. 4) by fixing a plane α. We then used α to introduce nonhomogeneous coordinates and then we averaged the metric ρ with respect to small translations. These translations form a subgroup of the projective group and, hence, the metric ρ_δ is obtained by averaging the metric ρ with respect to transformations A which are close to the identity transformation. If ϑ denotes the set of parameters of the transformation A, then the metric ρ_δ is defined by the formula

$$\rho_\delta(x, y) = \lambda \int_{|\vartheta| < \delta} \varphi(\vartheta)\, \rho(Ax, Ay)\, d\vartheta.$$

The average ρ_δ of the metric ρ makes sense in the whole space if it is interpreted in this way. Moreover, ρ_δ is a Desarguesian metric and, outside the plane α, it has the regularity properties established in Section 5. In the case of a metric of elliptic type, the metric ρ_δ converges to the metric ρ in the whole space as $\delta \to 0$.

The transition from the metric ρ_δ to the metric $\rho_{\delta\varepsilon}$ also represents an averaging and, hence, again reduces to a Desarguesian metric, converging uniformly to the metric ρ_δ in the whole space as $\varepsilon \to 0$. The transition from the metric $\rho_{\delta\varepsilon}$ to the metric $\rho_{\delta\varepsilon\alpha}$ is also an averaging of the metric $\rho_{\delta\varepsilon}$, but this time with respect to another subgroup. The metric $\rho_{\delta\varepsilon\alpha}$ converges uniformly to the metric $\rho_{\delta\varepsilon}$ as $\alpha \to 0$, and is regular (of class C^∞) everywhere with the possible exception of the special plane α.

If we now take another plane α_1 and apply the same system of averagings to the metric $\rho_{\delta\varepsilon\alpha}$, we get a Desarguesian metric which is regular everywhere with the possible exception of the line g in which the plane α_1 intersects α. We can then take a plane α_2 and use the described system of averagings to obtain a Desarguesian metric which is regular everywhere with the possible exception of the point A in which the plane α_2 intersects the line g. Finally, if we take a plane α_3 that does not pass through the point A and carry out the averagings, we obtain a Desarguesian metric which is regular in the whole space. Moreover, if the parameters δ, ε, α, $\ldots$ are appropriately small, then the resulting regular Desarguesian metric will be arbitrarily close to the original metric. The theorem is now completely proved.

We can obtain a Desarguesian metric approximating the given general Desarguesian metric in a single step, by averaging over the transformations of a subgroup with a sufficiently large number of parameters. In fact, if we take the subgroup of elliptic motions, then a Desarguesian metric which approximates the given metric and has regularity of class C^∞, can be obtained by using the formula

$$\overline{\rho}_\delta(x, y) = \lambda \int\limits_{|\vartheta| < \delta} \varphi(\vartheta)\rho(Ax, Ay)\,d\vartheta,$$

where A is the transformation of the subgroup of elliptic motions corresponding to the set of parameters ϑ. Here it is assumed that $\vartheta = 0$ corresponds to the identity transformation.

§7. GENERAL DESARGUESIAN METRICS IN THE TWO-DIMENSIONAL CASE

In the present section we will obtain a basic result for two-dimensional Desarguesian metrics, namely, we will prove that they are all σ-metrics. The concept of a σ-metric in the projective plane is introduced in the same way as in space, except that here lines play the roles of planes. Let us recall the definition of a σ-metric, as applied to the two-dimensional case.

Let X be any set of points of the plane and let gX denote the set of lines intersecting X. Moreover, let σ be any nonnegative completely additive set function, defined on the sets of lines of the projective plane, which satisfies the following conditions: 1) σ $(gX) = 0$ if the set X consists of a single point; 2) σ $(gX) > 0$ if X contains a line interval. Using the function σ, we define the length of a line interval as the value of the function σ on the set of lines intersecting the interval. We then define the metric ρ (x, y), namely, the *distance between two points x* and *y,* to be the length of the shorter of the two intervals joining the points. Just as in the case of space, we prove that the concept of distance introduced in this way satisfies the axioms of a metric space and then prove that the metric ρ (x, y) is Desarguesian.

Theorem. *Every complete continuous two-dimensional Desarguesian metric is a σ-metric.*

First, we consider two-dimensional metrics of the hyperbolic and parabolic types. In this case, we can regard the domain of definition G_0 of the metric as being the Euclidean plane or a bounded convex domain in the Euclidean plane. As shown in Section 6, a general Desarguesian metric can be approximated by a regular Desarguesian metric converging uniformly to the given metric on every compact subset of G_0. It was proved in Section 4 that a regular two-dimensional Desarguesian metric is specified by a line element

$$ds = \int_\omega \gamma \, (g) \; |\xi \, dx| \; d\xi,$$

where g is the line passing through the point x perpendicular to the unit vector ξ, $\gamma(g)$ is a regular nonnegative function defined on the set of lines, and the integration is extended over the circumference of the unit circle with center at the point x. We use the density $\gamma(g)$ to define a completely additive function σ, by setting

$$d\sigma = \gamma(g)\,dg,$$

where $dg = dp\,d\xi$, and p, ξ are the coefficients in the equation of the line $x\xi - p = 0$. Using the function σ, we can rewrite the line element ds of the regular Desarguesian metric in the form

$$ds = \int \gamma(g)\,dp\,d\xi = \int_{g\,\cap\,dx\,\neq\,0} \gamma(g)\,dg,$$

where the integration on the right is extended over the set of lines g that intersect the element dx. It follows that the length of an arbitrary curve c in the Desarguesian metric under consideration is given by

$$L(c) = \int_{g\,\cap\,c\,\neq\,0} n\gamma(g)\,dg,$$

where $n = n(g, c)$ is the number of points of intersection of the line g with the curve c, and the integration is extended over the set of lines intersecting the curve. In the case of a straight line interval δ, we have $n = 1$, and hence

$$L(\delta) = \int_{\delta\,\cap\,g\,\neq\,0} \gamma(g)\,ds = \sigma(g\delta),$$

where $\sigma(g\delta)$ is the value of the set function σ on the set of lines intersecting the segment δ.

Our constructions involve completely additive functions on the sets of lines. In order to make these constructions more intuitive, we use a correlation transformation to go from functions on the set of lines to functions on the set of points. This is done by assigning to the line

$$ax_1 + bx_2 = 1$$

the point with Cartesian coordinates $x_1 = a$, $x_2 = b$. In this way, we establish a one-to-one correspondence between the lines and the points of a plane punctured at the origin of coordinates, the lines of a pencil being assigned points on a straight line.

Let σ be a completely additive set function defined on the sets of lines, where σ specifies a Desarguesian metric. We define a completely additive function σ' on the sets of points of the plane by putting $\sigma'(M') = \sigma(M)$, where M is the set of lines corresponding to the set of points M' under the correlation transformation. In this way, we go over from the function σ defined on sets of lines to a function σ' defined on point sets. The fact that lines passing through the origin of coordinates have no point images is of no significance for us, since the set functions σ considered here vanish on the set of lines belonging to any pencil.

We now explain the meaning of the values of the function σ' for certain special sets. To be explicit, suppose the metric defined by the function σ is of hyperbolic type. In this case, the domain of definition of the function σ is the sets of lines intersecting a bounded convex domain G. If the origin of coordinates is chosen inside the domain G, then the domain of definition of the function σ' will be the set of points lying outside some finite convex domain G'.

Let g be an arbitrary line that does not intersect the closure of the domain G' and, of the two half-planes determined by the line g, let $\bar{\alpha}_g$ be the one that does not contain G'. Under the correlation, the set of points of the half-plane $\bar{\alpha}_g$ is assigned to a set of lines intersecting some segment δ lying strictly inside G. In fact, let

$$ax_1 + bx_2 = 1$$

be the equation of the line g. Then we have

$$ax_1' + bx_2' \geq 1$$

for every point (x'_1, x'_2) of the half plane $\bar{\alpha}_g$. The line corresponding to the point (x'_1, x'_2) is given by the equation

$$x_1 x'_1 + x_2 x'_2 = 1.$$

Every such line intersects the segment δ with endpoints $(0, 0)$ and (a, b), since

$$0x'_1 + 0x'_2 < 1, \text{while } ax'_1 + bx'_2 \geqslant 1.$$

Conversely, under the correlation, every line $x_1 x'_1 + x_2 x'_2 = 1$ intersecting the segment δ is assigned a point of the half-plane $\bar{\alpha}_g$. Since the set of points of the half-plane $\bar{\alpha}_g$ is assigned the set of lines intersecting the segment δ, the quantity $\sigma'(\bar{\alpha}_g)$ is just the length of the segment δ in the metric defined by the function σ.

Similarly, it can be proved that the value of the function σ' on the set of points of a strip between parallel lines that do not intersect the closure of the domain G' is the length of a segment in the domain G. In fact, if the strip is bounded by the lines

$$ax_1 + bx_2 = 1, \quad \lambda a x_1 + \lambda b x_2 = 1,$$

then the segment we are talking about is just the one joining the points (a, b) and $(\lambda a, \lambda b)$.

Let G'' be a convex domain which contains the domain G' together with its boundary, and let $\bar{G}''$ denote the set of points lying outside the domain G''. Then the value of the function σ' on the whole set of points of $\bar{G}''$ is finite, and the value of σ' on an ε-neighborhood of an arbitrary point of $\bar{G}''$ is small whenever ε is small. In fact, the set $\bar{G}''$ can be covered by a finite number of half-planes which do not intersect the closure of the domain G'. The value of σ' is finite for every half-plane, and hence the value of σ' is also finite for the whole set $\bar{G}''$. Moreover, an ε-neighborhood of any point can be included in an ε-strip in $\bar{G}''$, and the value of the function σ' on an ε-neighborhood is no larger than its value on the strip containing the ε-neighborhood. But the value of the function σ' on the strip is given by the length of the segment corresponding to the strip under the correlation. Since the width of the strip is small, the segment corresponding to it is small and hence has a small length, by the continuity of the metric.

This result, together with its proof, carries over to the case of an infinite sequence of σ-metrics which converges uniformly in every compact subdomain of a domain G to a Desarguesian metric defined in G. Namely, if σ_n is a sequence of completely additive functions whose corresponding σ-metrics converge to a continuous Desarguesian metric in G, then the values of the σ'_n on the set $\overline{G}''$ are uniformly bounded, and the values of the σ'_n on ε-neighborhoods of points of $\overline{G}''$ are uniformly small. The first assertion can be strengthened, i.e., it can be proved that the values $\sigma_n\,(\tilde{G}'')$ converge uniformly (in n) to zero, when the domain G'' is enlarged without limit until it covers the whole plane.

We now show that *every complete continuous Desarguesian metric* $\rho\ (x,\ y)$ *is a σ-metric*. First, we recall the notation that has been introduced: G is the bounded convex domain in which the metric ρ is defined, G' is the convex domain whose exterior contains the images under correlation of the lines intersecting the domain G, $\overline{G}'$ is the complement of this domain, G'' is a convex domain whose interior contains the closure of the domain G', and $\overline{G}''$ is its complement. Since the function σ' is essentially the function σ, referred to point sets under a correlation of the plane, we will subsequently identify the function σ' with σ. In other words, in talking about the function σ on a point set, we have in mind the function σ'.

According to the theorem of Section 6, there exists a sequence of regular Desarguesian metrics $\rho_n\ (x,\ y)$ converging uniformly to the metric $\rho\ (x,y)$ in every compact subdomain of the domain G. The $\rho_n\ (x,y)$ are all σ-metrics. Let σ_n be the completely additive function corresponding to the metric ρ_n. First we decompose the plane into unit squares by drawing the lines $x_1 = n, x_2 = m$, where m and n are integers, then we decompose each of these squares into four squares of side $1/2$ by drawing lines parallel to the x_1 and x_2 axes, then we subject each of the new squares to an analogous decomposition into squares of side $1/4$, and so on. There are no more than countably many squares in this decomposition of the plane, and hence the squares can be enumerated. Let c be a square of the decomposition which lies strictly inside $\overline{G}'$. Since the values σ_n (c) are bounded, we can select a convergent sequence from them. Then, in the familiar fashion, we select a subsequence of the functions σ_n such that the numerical sequences σ_n (c) converge for every square c lying strictly inside $\overline{G}'$.

We now define a set function σ in $\overline{G}'$ as follows. Given any square c of the decomposition of the plane constructed above which lies strictly inside $\overline{G}'$, we set

$$\sigma (c) = \lim_{n \to \infty} \sigma_n (c).$$

To define the value of the function σ for any open set $B \subset \overline{G}'$, let M_1 be the set of unit squares of the decomposition contained in B, let M_2 be the set of squares of side $1/2$ contained in $B - M_1$, let M_3 be the set of squares of side $1/4$ contained in $B - M_1 - M_2$, and so on. We then write

$$\sigma (B) = \sum_k \sum_{c \subset M_k} \sigma (c).$$

For an arbitrary set A we write

$$\sigma (A) = \inf_{A \subset B} \sigma (B),$$

where the infimum is taken over all open sets containing A.

Next, we show that the set function σ constructed in this way is a nonnegative completely additive function on the ring of Borel sets. As is well-known, the following two conditions are sufficient for a function to be completely additive on the ring of Borel sets:

1) $\sigma (A_1 \cup A_2) = \sigma (A_1) + \sigma (A_2)$ for every pair of sets A_1, A_2 whose closures do not intersect;

2) $\sigma (\bigcup_k A_k) \leqslant \sum_k \sigma (A_k)$ for arbitrary sets $A_1, A_2, A_3, \ldots$.

Let A_1 and A_2 be any two sets whose closures do not intersect.

Then they satisfy the first condition. To see this, we note that since the closures of the sets A_1 and A_2 do not intersect, there exist nonintersecting open sets B_1 and B_2 containing A_1 and A_2, respectively. According to the definition of σ, there exist open sets B_1', B_2' and B', containing A_1, A_2 and $A_1 \cup A_2$, respectively, such that

$$\sigma(B_1') - \sigma(A_1) < \varepsilon, \quad \sigma(B_2') - \sigma(A_2) < \varepsilon,$$
$$\sigma(B) - \sigma(A_1 \cup A_2) < \varepsilon.$$

Let $B_1'' = B_1' \cap B_1 \cap B$, $B_2'' = B_2' \cap B_2 \cap B$, $B'' = B_1'' \cup B_2''$. The sets B_1'' and B_2'' do not intersect. It is an immediate consequence of the definition of the function σ for open sets that

$$\sigma(B_1'') + \sigma(B_2'') = \sigma(B'').$$

On the other hand, by the choice of the sets B_1', B_2' and B',

$$\sigma(B_1'') - \sigma(A_1) < \varepsilon, \quad \sigma(R_2'') - \sigma(A_2) < \varepsilon,$$
$$\sigma(B) - \sigma(A_1 \cup A_2) < \varepsilon,$$

which implies

$$|\sigma(A_1) + \sigma(A_2) - \sigma(A_1 \cup A_2)| < 3\varepsilon.$$

Therefore, since the choice of ε is arbitrary,

$$\sigma(A_1) + \sigma(A_2) = \sigma(A_1 \cup A_2),$$

i.e., the function σ satisfies the first condition.

We now verify the validity of the second condition. By the definition of σ, for every set A_k there exists an open set B_k containing A_k such that

$$\sigma(B_k) - \sigma(A_k) < \varepsilon/2^k.$$

For open sets B_k it is obvious that

$$\sigma\left(\bigcup_h B_k \right) \leqslant \sum_h \sigma(B_k)$$

and hence

$$\sigma\left(\bigcup_h A_k \right) \leqslant \sum_h \sigma(A_k) + \varepsilon.$$

But the choice of ε was arbitrary and, hence,

$$\sigma\left(\bigcup_h A_k \right) \leqslant \sum_h \sigma(A_k),$$

so that the function σ also satisfies the second condition. Thus the function σ' is a nonnegative completely additive function on the Borel sets in $\overline{G}'$.

Finally we show that the σ-metric $\bar{\rho}\,(x,\,y)$, defined with the help of the function σ just constructed, coincides with the metric $\rho\,(x,\,y)$, the limit of the metrics $\rho_n\,(x,\,y)$ defined by the functions σ_n. Let δ be an arbitrary segment inside the domain G, whose endpoints have coordinates $x'_1,\ x'_2$ and $x''_1,\ x''_2$. Its length in the metric $\bar{\rho}$ is given by the value of the function σ on the intersection X of the half-planes satisfying the inequality

$$(x_1 x'_1 + x_2 x'_2)^{-1}\,(x_1 x''_1 + x_2 x''_2)^{-1} \leqslant 0.$$

The length of the segment δ in the metric ρ_n is the value of the function σ_n on the intersection of the same half-planes. Thus we must prove that $\sigma(X) = \lim \sigma_n(X)$ as $n \to \infty$.

Let X_α denote the set of points in X whose distance from the boundary of X is no less than α ($\alpha > 0$) and whose distance from the origin of coordinates is no greater than $1/\alpha$. The set X_α is bounded and lies strictly inside X. By what was proved earlier, given any $\varepsilon > 0$, we have $\sigma_n(X - X_\alpha) < \varepsilon$ for sufficiently small α and sufficiently large n. From the decomposition of the plane constructed above we now take a finite system of nonintersecting squares c_k which are contained in X and cover X_α. For this system of squares

$\sigma_n(X) - \sigma_n(\cup c_k) < \varepsilon$, since $\sigma_n(X - X_\alpha) < \varepsilon$. On the other hand, by the definition of σ,

$$\sigma(X) - \sigma_n(\cup c_k) < \varepsilon$$

for sufficiently large n. It follows from these inequalities that

$$|\sigma(X) - \sigma_n(X)| < 2\varepsilon$$

for sufficiently large n. Therefore $\rho_n \to \bar{\rho}$, i.e., $\rho = \bar{\rho}$. Thus the theorem is proved for metrics of hyperbolic type.

The proof is analogous for metrics of parabolic type. In the case of metrics $\rho(x, y)$ of elliptic type, the same argument leads to the conclusion that the metric $\bar{\rho}$ coincides with the metric ρ on the projective plane cut along the line at infinity. But since both metrics are continuous, they coincide on the whole plane.

§8. FUNK'S PROBLEM

Given a convex body[14] T in three-dimensional Euclidean space, we introduce rectangular Cartesian coordinates x_1, x_2, x_3 with the origin of coordinates inside T. By the *support function* of the body T we mean the function $F(x)$ defined in all of space by the equation

$$F(x) = \sup_{y \subset T} xy,$$

where the supremum is taken over all points y of the body T. The function F is positive homogeneous of degree 1 and convex. In fact,

$$F(\lambda x' + \mu x'') = \sup_{y \in T} y(\lambda x' + \mu x'')$$

$$\leqslant \lambda \sup_{y \in T} yx' + \mu \sup_{y \in T} yx'' = \lambda F(x') + \mu F(x'')$$

for arbitrary $\lambda, \mu \geqslant 0$. Conversely, if a function $F(x)$ is positive homogeneous of degree 1 and convex, then it is the support function of some convex body. This convex body is the intersection of the half-spaces E_y defined by the inequalities

$$xy \leqslant F(y).$$

Let $\gamma\,(\xi)$ be a nonnegative even function of the unit vector ξ, defined on the sphere ω: $|\xi| = 1$. Then the function

$$F(x) = \int_\omega \lambda(\xi)\,|x\xi|\,d\xi \tag{1}$$

is even in x, convex and positive homogeneous of degree 1. The convex body with $F(x)$ as its support function has its center of symmetry at the origin of coordinates. *The problem of representing the support function of an arbitrary centrally symmetric convex body in the form* (1), *with an even but not necessarily positive function* $\gamma\,(\xi)$, *is known as Funk's problem.* This problem is of interest to us in connection with Desarguesian metrics in three-dimensional space, which will be considered in the next section.

We will reduce Funk's problem to that of reconstructing a continuous even function, defined on the unit sphere ω, from its average values on the great circles of the sphere. We begin by proving the following identity:

$$\int_\omega \bar{\psi}\,(\xi)\,d\xi - \int_{(\xi\eta)^2 > \lambda^2} \frac{\bar{\psi}(\xi)\,|\xi\eta|\,d\xi}{\sqrt{(\xi\eta)^2 - \lambda^2}} = \int_{(\xi\eta)^2 > 1-\lambda^2} \psi\,(\xi)\,d\xi. \tag{2}$$

Here $\psi\,(\xi)$ is a continuous even function defined on the sphere ω, and $\bar{\psi}\,(\xi)$ is the average value of the function ψ on the great circle with pole ξ.

To prove the identity (2), we first prove that it holds to within an accuracy of ε^2 for a function $\psi\,(\xi)$, which equals 1 in an ε-neighborhood of an arbitrary point ζ and of the point $-\zeta$ antipodal to ζ, and vanishes outside these neighborhoods. We assume that ε is sufficiently small and that $(\zeta\eta)^2 \neq 1 - \lambda^2$. We will consider two cases depending on the relative position of the points η and ζ namely, *Case 1*: $(\eta\zeta)^2 > 1 - \lambda^2$, and *Case 2*: $(\eta\zeta)^2 < 1 - \lambda^2$. First we evaluate the integral

$$J_1 = \int\limits_{(\xi\eta)^2 > 1 - \lambda^2} \psi\ (\xi)\ d\xi.$$

It is obvious that in the second case, the ε-neighborhood of the point ζ lies outside the region of integration if ε is sufficiently small, and hence $J_1 = 0$. In the first case, the ε-neighborhood of ζ belongs entirely to the region of integration if ε is sufficiently small, and hence $J_1 = 2\pi\varepsilon^2$ up to quantities of higher order in ε.

We now consider the integral

$$J_2 = \int\limits_\omega \overline{\psi}\ (\xi)\ d\xi.$$

The function $\overline{\psi}\ (\xi)$ is different from zero only in an ε-neighborhood of the great circle $\varkappa(\zeta)$ with pole at the point ζ. If ε is sufficiently small, the value of the function $\overline{\psi} \simeq b\ (\delta)/2\pi$ at the distance δ from $\varkappa\ (\zeta)$, where $b(\delta)$ is the chord of the ε-neighborhood $N_\varepsilon\ (\zeta)$ of the point ζ at the distance δ from its center. In evaluating J_2, we divide the integration over the ε-neighborhood of the circle $\varkappa\ (\zeta)$ into an integration in the direction perpendicular to the circle $\varkappa\ (\zeta)$ and an integration in the direction of $\varkappa\ (\zeta)$. The first integration obviously gives the area of a disk of radius ε, divided by π, i.e., ε^2, while the second integration, apart from terms of higher order in ε, reduces to multiplication by the length of $\varkappa\ (\zeta)$, i.e., by 2π. Therefore, up to terms of order ε^2,

$$J_2 = 2\pi\varepsilon^2.$$

Finally we consider the integral

$$J_3 = \int\limits_{(\xi\eta)^2 > \lambda^2} \frac{\overline{\psi}\ (\xi)\ |\xi\eta|\ d\xi}{\sqrt{(\xi\eta)^2 - \lambda^2}}.$$

Just as for the integral J_1, we will distinguish two cases, depending on the relative position of the points ζ and η, namely, *Case 1*: $(\eta\zeta)^2 > 1 - \lambda^2$, and *Case 2*: $(\eta\zeta)^2 < 1 - \lambda^2$.

In the first case, the ε-neighborhood of the circle $\varkappa(\zeta)$ lies outside the region of integration $(\xi\eta)^2 > \lambda^2$ if ε is sufficiently small, and hence the integral J_3 vanishes. In the second case, i.e., when $(\eta\zeta)^2 < 1 - \lambda^2$, the ε-neighborhood of the circle $\varkappa(\zeta)$ intersects the region of integration. To evaluate the integral J_3, we use the same argument as in the case of the integral J_2. First we integrate in the direction perpendicular to $\varkappa(\zeta)$, obtaining

$$\varepsilon^2 \frac{|\xi\eta|}{\sqrt{(\xi\eta)^2 - \lambda^2}},$$

and then we integrate along $\varkappa(\zeta)$ to get the value of the integral

$$J_3 = \varepsilon^2 \int_{(\xi\eta)^2 > \lambda^2} \frac{|\xi\eta|\, ds}{\sqrt{(\xi\eta)^2 - \lambda^2}},$$

where the variable of integration is the arclength s of the circle $\varkappa(\zeta)$. To evaluate this integral, we introduce rectangular Cartesian coordinates x, y, z with the origin of coordinates at the center of the sphere ω, the z axis directed at the point η, and the x axis perpendicular to the vectors η, ζ. We then obtain

$$\int_{(\xi\eta)^2 > \lambda^2} \frac{|\xi\eta|\, ds}{\sqrt{(\xi\eta)^2 - \lambda^2}} = 4 \int_\lambda^\mu \frac{dz}{\sqrt{\mu^2 - z^2}\sqrt{z^2 - \lambda^2}},$$

where $\mu^2 = 1 - (\eta\zeta)^2$. The integral on the right has the same value $\pi/2$ for arbitrary λ and μ ($\lambda < \mu$). Therefore in the second case, the integral J_3 is equal to

$$J_3 = 2\pi\varepsilon^2.$$

Thus we see that equation (2) holds up to quantities of order ε^2 in both cases, i.e., for both relative positions of the points η and ζ.

We now take an arbitrary continuous even function ψ on the sphere ω. Suppose equation (2) does not hold for such a function. The function ψ can be approximated by a linear combination of functions ψ_ζ in such a way that the difference

$$\left| \psi\,(\xi) - \sum_h \psi\,(\zeta_h)\ \psi_{\zeta_h}\,(\xi) \right|$$

equals $|\psi\,(\xi)|$ on a set of measure less than δ, while the difference is less than δ on the remaining part of the sphere ω. Equation (2) holds for a linear combination of the function ψ_ζ with any preassigned accuracy, if ε is sufficiently small. But then, contrary to hypothesis, (2) holds for the function ψ with any preassigned accuracy, if δ is sufficiently small, and the identity is proved.

Dividing the identity (2) by $\pi\lambda^2$ and taking the limit as $\lambda^2 \to 0$, we get a representation of the function ψ in terms of its average values on great circles, namely

$$\psi\,(\eta) = \frac{1}{\pi}\,\frac{d}{d\,(\lambda^2)} \int_{(\xi\eta)^2 > \lambda^2} \frac{\overline{\psi}\,(\xi)\,|\xi\eta|\,d\xi}{\sqrt{(\xi\eta)^2 - \lambda^2}} \Bigg|_{\lambda^2 = 0}, \qquad (3)$$

We now return to Funk's problem. Let $F\,(x)$ be the support function of a central symmetric convex body T, and suppose $F\,(x)$ has a representation of the form (1) with a continuous even function γ. Let us find the sum of the principal radii of curvature of the surface of the body T at the point with exterior normal η. As is well known

$$R_1 + R_2 = \Delta F\,(x)\,\big|_{x=\eta},$$

where Δ is the Laplace operator (see [2], [3], or [4]). We introduce rectangular Cartesian coordinates, taking the direction of η to be the direction of the x_3 axis. Then $\partial^2 F/\partial x_3^2 = 0$ at the point η, by the homogeneity of the function F and, hence,

$$R_1 + R_2 = \left(\frac{\partial^2 F}{\partial x_1^2} + \frac{\partial^2 F}{\partial x_2^2} \right)\Big|_\eta \, .$$

Next we calculate the derivatives $\partial^2 F/\partial x_1^2$, $\partial^2 F/\partial x_2^2$ at the point η, starting from the representation (1) for the function F. Since the function $\gamma\,(\xi)$ is even,

$$\frac{\partial F}{\partial x_1} = 2 \int\limits_{x\xi<0} \lambda(\xi)\,\xi_1\,d\xi$$

for arbitrary x. To calculate the derivative $\partial^2 F/\partial x_1^2$, we first note that

$$\frac{\partial F}{\partial x_1}\Big|_{x+dx} - \frac{\partial F}{\partial x_1}\Big|_{x} = 2\int\limits_{X} \gamma\,(\xi)\,\xi_1\,d\xi,$$

where the integration is carried out over the region X between the two neighboring circles with poles $\eta + d\eta$ and η. Since the circles are neighboring, $d\xi = \xi_1\,ds$, where ds is the element of arclength of the circle with pole η. Taking the limit as $dx_1 \to 0$, we get

$$\frac{\partial^2 F}{\partial x_1^2} = 2 \int\limits_{\xi\eta=0} \gamma\,(\xi)\,\xi_1^2\,ds,$$

where the integration is carried out along the arc of the great circle with pole η. Similarly, we have

$$\frac{\partial^2 F}{\partial x_2^2} = 2 \int\limits_{\xi\eta=0} \gamma\,(\xi)\,\xi_2^2\,ds.$$

Moreover,

$$\left(\frac{\partial^2 F}{\partial x_1^2} + \frac{\partial^2 F}{\partial x_2^2}\right)_\eta = 2 \int\limits_{\xi\eta=0} \gamma\,(\xi)\, ds,$$

since $\xi_1^2 + \xi_2^2 = 1$ if $\xi\eta = 0$ and, therefore,

$$(R_1 + R_2)_\eta = 2 \int\limits_{\xi\eta=0} \gamma\,(\xi)\, ds.$$

Now let $F(x)$ be any convex function, which is even in x and three times differentiable. Then $F(x)$ can be represented in the form (1). To prove this, we consider the integral equation

$$\Delta F\,(x)\,|_\eta = 2 \int\limits_{\xi\eta=0} \gamma\,(\xi)\, ds \tag{4}$$

for the function $\gamma(\xi)$. This equation has a unique solution (formula (3)). Now consider the two functions

$$F_1\,(x) = \int\limits_\omega \gamma\,(\xi)\,|x\xi|\,d\xi + R\,|x|,$$
$$F_2\,(x) = F\,(x) + R\,|x|.$$

Both of these functions are convex, for sufficiently large R, and hence are support functions of convex bodies T_1 and T_2. Calculating the sum of the principal radii of curvature of the surfaces of these bodies at the points with exterior normal η, we get

$$(R_1 + R_2)_{T_1} = \Delta F_1\,|_\eta = 2 \int\limits_{\xi\eta=0} \gamma\,(\xi)\, ds + 2R,$$
$$(R_1 + R_2)_{T_2} = \Delta F_2|_\eta = \Delta F|_\eta + 2R.$$

Since $\gamma\,(\xi)$ satisfies equation (4), we have

$$(R_1 + R_2)_{T_1} = (R_1 + R_2)_{T_2},$$

i.e., the sums of the principal radii of curvature of the surfaces of the bodies T_1 and T_2 at points with parallel exterior normals are equal, and hence, by Christoffel's Theorem (see [3]), T_1 and T_2 are equal to within a parallel translation. But then, since T_1 and T_2 are symmetric with respect to the origin of coordinates, they must coincide completely and hence so do their support functions $F_1(x)$ and $F_2(x)$. It follows that

$$F(x) = \int_\omega \gamma(\xi) |x\xi| \, d\xi,$$

as was to be proven. The formula

$$\lambda(\eta) = \frac{1}{\pi^2} \frac{d}{d(\lambda^2)} \int_{(\xi\eta)^2 > \lambda^2} \frac{\Delta F|_\xi \, |\xi\eta| \, d\xi}{\sqrt{(\xi\eta)^2 - \lambda^2}} \Big|_{\lambda=0} \tag{5}$$

expresses the function $\gamma(\xi)$ in terms of the function $F(x)$.

§9. DESARGUESIAN METRICS IN THE THREE-DIMENSIONAL CASE

In this section we shall give a constructive description of general Desarguesian metrics in the three-dimensional case. As shown in Section 6, a general Desarguesian metric can be approximated by regular Desarguesian metrics converging uniformly in every compact subset of the domain of definition of the metric. We first find the general representation of regular Desarguesian metrics. Let

$$ds = F(x, dx)$$

be the line element of a regular Desarguesian metric. The function

$F(x, \xi)$ is even and convex in ξ and, as shown in Section 8, has the representation

$$F(x, \dot{x}) = \int_\omega \gamma(x, \xi) |\dot{x}\xi| d\xi, \qquad (1)$$

where γ is a function even in ξ, and the integration is carried out over the unit sphere ω: $|\xi| = 1$.

Since the metric in question is Desarguesian, the extremals of the length functional

$$s = \int F(x, \dot{x}) dt$$

are straight lines. The Euler equations for the functional s are

$$\sum_{\alpha,\beta} \left[\frac{\partial F}{\partial x_h} - \dot{x}_\alpha \frac{\partial^2 F}{\partial \dot{x}_h \partial x_\alpha} - \ddot{x}_\beta \frac{\partial^2 F}{\partial \dot{x}_h \partial \dot{x}_\beta} \right] = 0$$

Choosing the Euclidean arc length along the extremal as the parameter t, we have $\ddot{x}_\beta = 0$ and therefore

$$\sum_\alpha \left[\frac{\partial F}{\partial x_h} - \dot{x}_\alpha \frac{\partial^2 F}{\partial \dot{x}_h \partial x_\alpha} \right] = 0 \qquad (2)$$

Since an extremal (a straight line) passes through every point x in every direction, equation (2) must hold for arbitrary x and $\dot{x}$. Let us find the general form of the function F satisfying equation (2).

Setting

$$\Phi = \sum_\beta \dot{x}_\beta \frac{\partial F}{\partial x_\beta},$$

we can rewrite equation (2) as

$$2 \frac{\partial F}{\partial x_k} - \frac{\partial \Phi}{\partial \dot{x}_k} = 0.$$

It follows that

$$\frac{\partial^2 F}{\partial x_\alpha \, \partial \dot{x}_\beta} - \frac{\partial^2 F}{\partial x_\beta \, \partial \dot{x}_\alpha} = 0 \tag{3}$$

for arbitrary α and β. To find the derivatives $\partial^2 F / \partial x_\alpha \, \partial \dot{x}_\beta$ and $\partial^2 F / \partial x_\beta \, \partial \dot{x}_\alpha$, we start from the representation (1) of the function F. Differentiating formula (1) first with respect to x_α and then with respect to $\dot{x}_\beta$, we get

$$\frac{\partial F}{\partial x_\alpha} = \int_\omega \frac{\partial \gamma}{\partial x_\alpha} |\dot{x}\xi| \, d\xi,$$

$$\frac{\partial^2 F}{\partial x_\alpha \, \partial \dot{x}_\beta} = 2 \int_{\dot{x}\xi > 0} \frac{\partial \lambda}{\partial x_\alpha} \xi_\beta \, d\xi,$$

where the integration on the right is extended over the hemisphere on which $\dot{x}\xi > 0$. Similarly,

$$\frac{\partial^2 F}{\partial x_\beta \, \partial \dot{x}_\alpha} = 2 \int_{\dot{x}\xi > 0} \frac{\partial \gamma}{\partial x_\beta} \xi_\alpha \, d\xi.$$

Substituting these values of the derivatives into formula (3), we obtain

$$\int_{\dot{x}\xi > 0} \left(\frac{\partial \gamma}{\partial x_\alpha} \xi_\beta - \frac{\partial \gamma}{\partial x_\beta} \xi_\alpha \right) d\xi = 0,$$

or

$$\int_{\dot{x}\xi > 0} (\nabla \gamma \times \xi) \, d\xi = 0 \tag{4}$$

in vector form, where ∇ is the gradient symbol and $\times$ denotes the vector cross product.

Next we show that equation (4) implies that $\nabla \gamma \times \xi \equiv 0$. Let η be the unit vector of the direction $\dot{x}$, and let η_0 be the unit vector of the direction of the x_3 axis. We have

$$\int_{\eta_0\xi > 0} (\nabla \gamma \times \xi) \, d\xi = 0, \qquad \int_{\eta\xi > 0} (\nabla \gamma \times \xi) \, d\xi = 0.$$

Subtracting these equations term for term, and noting that the function $\nabla \gamma \times \xi$ is even in ξ, we conclude that

$$\int_{\omega_\eta} (\nabla \gamma \times \xi) \, d\xi = 0, \tag{5}$$

where ω_η is the region cut out of the hemisphere $\eta_0\xi > 0$ by the great circle $\eta\xi = 0$. Without loss of generality, we can assume that the point x coincides with the origin of coordinates $(x_1, x_2, x_3 = 0)$. Projecting the region ω_η from the origin of coordinates onto the plane $x_3 = 1$, we get some half-plane E_η. Then, transforming in equation (5) from integration over the region ω_η to integration over the half-plane E_η, we find that

$$\int_{E_\eta} (\nabla \gamma \times \xi) \, \lambda \, dx_1 \, dx_2 = 0. \tag{6}$$

As $x_1^2 + x_2^2 \to \infty$, the function λ falls off like $1/(x_1^2 + x_2^2)^{3/2}$. The function $(\nabla \gamma \times \xi) \, \lambda$, as a function of x_1 and x_2, is summable on the $x_1 x_2$ plane. As is well-known, a summable continuous function is uniquely determined by giving its integrals on all half-planes. It follows that $(\nabla \gamma \times \xi)\lambda \equiv 0$. But $\lambda \neq 0$, and therefore $\nabla \gamma \times \xi = 0$, as asserted.

Since $\nabla \gamma \times \xi = 0$, the vector $\nabla \gamma$ is collinear with ξ and, hence, is perpendicular to any vector η perpendicular to ξ. It follows that $\nabla \gamma = 0$ for every such vector η. But this means that the derivative of the function γ with respect to any direction perpendicular to ξ equals zero, and hence γ has a constant value on the plane α passing through the point x perpendicular to the vector ξ. Thus, the line element of a regular Desarguesian metric is of the form

$$ds = \int_\omega \gamma(\alpha) \, |\xi \, dx| \, d\xi, \tag{7}$$

where γ is a function defined on planes, and α is the plane passing through the point x perpendicular to the vector ξ. If the plane is specified by an equation $x \cdot \xi - p = 0$ in normal form, then the line element ds can be written as

$$ds = \int_\omega \gamma(\xi, x\xi) \, |\xi \, dx| \, d\xi,$$

where γ is an even function in ξ.

Next we define a function σ on sets of planes in space, by writing

$$d\sigma = \gamma(\alpha) \, d\alpha = \gamma(\xi, x\xi) \, dp \, d\xi.$$

With the help of the function σ, we can write the line element ds in the form

$$ds = \int_\omega \gamma \, dp \, d\xi = \int_{\alpha \, \cap \, dx \neq 0} \gamma(\alpha) \, d\alpha = \sigma \big|_{\alpha \cap dx \neq 0}.$$

The length of an arbitrary segment δ is given by

$$L(\delta) = \int_{\delta \, \cap \, \alpha \neq 0} \gamma(\alpha) \, d\alpha = \sigma \big|_{\delta \cap \alpha \neq 0},$$

where the value of the function σ is taken on the set of planes intersecting the segment δ. Thus we see that a regular Desarguesian metric is a σ-metric in the three-dimensional case, just as in the two-dimensional case. However, whereas the function σ is nonnegative in the two-dimensional case, so far we have no grounds for drawing this conclusion in the three-dimensional case. Moreover, as we now show, one can give a simple example of a Desarguesian σ-metric for which the function γ, and hence the function σ as well, takes negative values.

Let $\gamma_1(\xi)$ be a regular function, which is even in ξ and equal to zero outside an ε-neighborhood of the point ξ_0, and which satisfies the condition $\gamma_1(\xi) \geqslant -2$, $\gamma_1(\xi_0) = -2$ in this neighborhood. Consider the metric specified by the line element

$$ds = F(x, dx) = \int_\omega (1 + \gamma_1(\xi)) \, |\xi \, dx| \, d\xi.$$

As shown in Section 8,

$$\frac{\partial^2 F}{\partial \dot{x}_1^2} = \int_{\dot{x}\xi = 0} (1 + \gamma_1(\xi)) \, \xi_1 \, ds,$$

where the integration is extended over the arc of the circle $\dot{x}\xi = 0$. From this it is clear that if ε is sufficiently small, then $\partial^2 F/\partial \dot{x}_1^2 > 0$ independently of the direction of differentiation, i.e., of the direction of $\dot{x}_1$. This means that F is convex in $\dot{x}$. But then the function F obviously defines a regular Desarguesian metric, since it is convex in $\dot{x}$ and has the form (7).

On the other hand, the function $\gamma(\alpha)$ cannot be arbitrary, if only because the length of a segment must be positive. But this will not be the case if we take γ to be a negative function.

We now find necessary and sufficient conditions for a regular metric specified by the function σ to be Desarguesian in the three-dimensional case. So that these conditions may appear more intuitive, we use a correlation transformation to go over from the function σ defined on sets of planes in space to a function defined on point sets, just as we did in the case of two-dimensional metrics.

Suppose we associate to the point a the plane specified by the equation $ax = 0$. Let M be an arbitrary set of planes, and let M' be the set of points corresponding to M. We define a function on point sets by setting its value on the set M' equal to $\sigma(M)$. To avoid the introduction of new notation, we will also use σ to denote this new function defined on point sets. The length of the segment with endpoints a and b represents the value of the function σ on the set of points bounded by the two planes $a \cdot x = 0$ and $b \cdot x = 0$. There are two such sets, corresponding to the fact that a pair of points on a projective line determines two segments.

We now take any projective triangle with vertices a, b, c. The length of any side of this triangle in the given plane metric is less than the sum of the lengths of the other two sides. The difference

$$\Delta = s\,(a,\,b) + s\,(a,\,c) - s\,(b,\,c)$$

can be expressed in terms of the function σ. In fact, this difference is the value of the function σ on the solid trihedral angle bounded by the three planes $a \cdot x = 0$, $b \cdot x = 0$ and $c \cdot x = 0$. It follows that the value of the function σ on every solid trihedral angle must be positive. But then, since the function γ specifying the set function σ is continuous, the value of the function σ is nonnegative for every cone and positive for a cone containing interior points. The value of σ is obviously zero on every plane. Thus, for the regular metric specified by the set function σ to be Desarguesian, it is necessary that the value of σ vanish on every plane and be nonnegative on every cone (positive if the cone contains interior points). Conversely, it is not hard to see that every completely additive function σ with these properties specifies a Desarguesian σ-metric.

By a σ-*metric in the generalized sense* we shall mean any metric which is specified by a completely additive set function satisfying the above conditions, and also any metric $\rho\,(x,\,y)$ which is the limit of such metrics with uniform convergence on every compact subset of the domain of definition of ρ.

Theorem. *Every complete continuous three-dimensional Desarguesian metric is a σ-metric in the generalized sense.*

In fact, every complete continuous Desarguesian metric can be approximated by regular Desarguesian metrics (by the theorem in

Section 6). But every regular Desarguesian metric is a σ-metric in the generalized sense. This proves the theorem.

§10. AXIOMS OF THE CLASSICAL GEOMETRIES

In speaking of the classical geometries, we have in mind Euclidean geometry, Lobachevskian geometry and Riemannian (or elliptic) geometry. Each of these geometries is introduced axiomatically. It is natural to divide the system of axioms underlying a geometry into groups, each defining the properties of relations of a particular kind (incidence, order, congruence) in the given geometry.

The system of axioms of Euclidean geometry consists of three groups.[15] The *first group of axioms* deals with the incidence properties of points of a line or a plane. The following assertions make up the first group of axioms: *Every line contains at least two points. There is one and only one line passing through any two points A and B. There are at least three noncollinear points. Every plane contains at least one point. There is one and only one plane passing through any three noncollinear points. If two points of a line belong to a plane, then the whole line belongs to the plane. If two planes have a point in common, then they have at least one more point in common. There are at least four noncoplanar points. Given a line g and a point A not on g, let π be the plane passing through g and A; then there is no more than one line in π that passes through A and does not intersect g* (the axiom of parallels).

As a consequence of the axioms of the first group we obtain the following theorems: Two lines lying in the same plane either have a single point in common, or they have no common points at all (i.e., they do not intersect). Two planes either have no common points, or they have a line in common but no common points other than the points of this line. A plane and a line that does not lie in the plane either have no common points, or they have a single point in common. There is one and only one plane passing through a line and a point not on the line, or through two intersecting lines.

The *second group of axioms* are the so-called *axioms of order*. The axioms of this group express *properties involving the relative positions of points on the line or in the plane.* It is assumed that there are two opposite directions on a line and that every pair of points A and B is in a known position with respect to each of these directions, as expressed by saying that one of the points *"precedes"* the other. This relation is denoted by the sign $<$, so that the expression *"A precedes B"* is written symbolically as $A < B$. The axioms of order require the indicated relation to have the following properties: *If $A < B$ in one direction, then $B < A$ in the opposite direction. In either one of the two directions, $A < B$ excludes $B < A$. In either one of the two directions, if $A < B$ and $B < C$, then $A < C$. In either one of the two directions, for every point B one can find points A and C such that $A < B < C$. If the points of a line are divided into two nonempty classes such that in one of the two directions each point of the first class precedes each point of the second class, then either there is a point in the second class that precedes all the other points of the second class, or there is a point in the first class that succeeds all the other points of the first class (Dedekind's axiom). A line g lying in a plane α divides this plane into two half-planes such that two points X and Y not on the line g belong to different half-planes if and only if the line XY intersects the line g in a point Z for which $X < Z < Y$ in one of the two directions.*

The concepts of *segment* and *half-plane* can be defined by using the axioms of order, and then appropriate theorems involving these concepts can be proved. Let g be a line, and let A be a point on g. Given a fixed direction on g, the point A divides g into two parts (half-lines); $X < A$ for every point X of one of these parts, while $A < X$ for every point of the other part. This division of the line into half-lines does not depend on the direction chosen on the line. Let A and B be two points on the line g. If the condition $A < C < B$ or $B < C < A$ holds for a point C of g, then we say that the point C lies *between* A and B. This property of a point lying between two given points is independent of the direction on the line. Consider the part of the line g, all of whose points lie between A and B. This part of g is called the *segment AB,* and the points A and B are called the *endpoints* of the segment.

The following theorems are consequences of the axioms of order and the incidence axioms: Given three points A, B, C on a line g, one and only one point lies between the other two. Every segment contains at least one point (other than its endpoints). If B is a point of the segment AC, then the segment AB is contained in the segment AC. If B is a point of the segment AC, then every point of the segment AC, other than B, belongs either to the segment AB or to the segment BC. If a line g intersects the side AB of a triangle ABC and does not pass through one of the vertices of the triangle, then the line intersects one and only one of the other two sides BC and AC (Axiom of Pasch). By a *triangle* we mean a figure which consists of three noncollinear points (the *vertices of the triangle*) and three segments (the *sides of the triangle*) joining these points in pairs.

The *third group of axioms* defines the *concept of congruence for segments and angles*. This group of axioms can be divided into two subgroups. In the case of the first subgroup (the axioms of linear congruence) we talk only about the congruence of segments, while in the case of the second subgroup, we talk about the congruence of angles or of segments and angles. The first subgroup contains the following axioms: *The segment AB is congruent to the segment BA. If the segment AB is congruent to segments CD and EF, then the segments CD and EF are congruent. If B and B_1 are points of the segments AC and $A_1 C_1$, respectively, then the congruences $AB = A_1 B_1$ and $BC = B_1 C_1$ imply the congruence $AC = A_1 C_1$. Given any point A on a line g and a segment CD, then for a given direction on g there is a point B such that $A < B$ and $AB = CD$.*

The congruence axioms of the second subgroup consist of assertions involving congruence of angles or of segments and angles. In particular, the following axiom is to be found among the axioms of this subgroup: *If the congruences $AB = A_1 B_1$, $AC = A_1 C_1$ and $\angle A = \angle A_1$ hold for two triangles ABC and $A_1 B_1 C_1$, then so does the congruence $\angle B = \angle B_1$.* The other axioms of this subgroup involve only congruence properties for angles.

The most important consequence of the axioms of order and linear congruence is the theorem on the introduction of measure (length) for segments. According to this theorem, to every segment we can uniquely assign a positive number μ, which has the following properties: Congruent segments have the same value of μ. If the point B lies on the segment AC between the points A and C, then $\mu(AC) =$

$\mu (AB) + \mu (BC)$. For a given segment $A_0 B_0$, chosen as the unit of length, we have $\mu(A_0 B_0) = 1$. The length of a segment is uniquely defined to within a constant factor, depending on the choice of the unit of measurement.

The system of axioms for Lobachevskian geometry differs from that for Euclidean geometry only in the axiom of parallels, which in the Lobachevskian case asserts that if g is a line and A a point not on g, then there are *two* lines that pass through A and do not intersect g. We note that all the consequences of the axioms of Euclidean geometry listed above are obtained without recourse to the axiom of parallels. Therefore they are true not only for Euclidean geometry, but also for Lobachevskian geometry.

The system of axioms for elliptic geometry also splits into three groups, i.e., axioms of incidence, axioms of order, and axioms of congruence. The *first group of axioms* (the *axioms of incidence*) differs from the corresponding group of axioms for Euclidean and Lobachevskian geometry in two respects. First of all, the axiom requiring the existence of two points on a line is replaced by an axiom requiring the existence of three points on a line. Second, the axiom of parallels is replaced by an axiom asserting that any two coplanar lines intersect.

The following theorems are consequences of the axioms of incidence for elliptic geometry: A line and a plane always have a point in common. Any two planes have a line in common. Any three planes have a point in common. The most important consequence of the axioms of incidence for elliptic geometry is Desargues' theorem on the perspective configuration of triangles. By a *triangle* we mean a figure which consists of three noncollinear points (the *vertices of the triangle*) and three lines (the *sides of the triangle*) joining these points in pairs. Two triangles ABC and $A_1 B_1 C_1$ are said to have a center of perspectivity S if the lines AA_1, BB_1 and CC_1 intersect in a point S, and an axis of perspectivity s if the sides AB and $A_1 B_1$, BC and $B_1 C_1$, AC and $A_1 C_1$ intersect on a line s. According to Desargues' theorem, if two triangles ABC and $A_1 B_1 C_1$ have a center of perspectivity, then they also have an axis of perspectivity and conversely (Fig. 2). It should be noted that Desargues' theorem is true in Euclidean and Lobachevskian geometry if it is assumed that

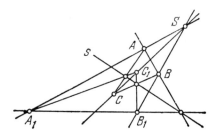

FIG. 2

the lines figuring in the definition of the center and axis of perspectivity have points of intersection.

In elliptic geometry we introduce the important concept of a harmonic quadruple of points. This concept is introduced with the help of a figure called a complete quadrangle. By a *complete quadrangle* we mean a figure made up of four points of the plane, no three of which are collinear, and six lines joining these points in pairs. Two sides of a complete quadrangle are said to be *opposite* if they have no vertices in common. The points of intersection of opposite sides are called *diagonal points*. We say that a pair of points C, D on a line is *harmonically conjugate* to a pair of points A, B if there exists a complete quadrangle for which the points A and B are diagonal, while the points C and D are the points of intersection of the line AB with the sides meeting at the third diagonal point (Fig. 3). The following theorems hold for harmonically conjugate pairs of points: Given a triple of points A, B, C, there is one and only one point D such that the pair C, D is harmonically conjugate to the pair A, B. The property of harmonic conjugacy of pairs is preserved under projection. The property of harmonic conjugacy of pairs is reciprocal. It is important to note that the concept of harmonically conjugate pairs of points can be introduced in Euclidean and Lobachevskian geometry if it is assumed that the lines figuring in the definition have points of intersection.

The *second group of axioms* of elliptic geometry consists of the

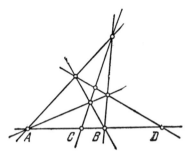

FIG. 3

axioms of order. It is assumed that there are two opposite directions on the line, and that with respect to each of these directions, every triple of points on the line stands in a certain relation, as expressed by saying that the points "succeed" each other. The relation of "succession" satisfies the following axioms: *The succession of three points A, B, C in one direction excludes their succession in the opposite direction. If three points A, B, C succeed each other in one direction, then the triples of points B, A, C and A, C, B succeed each other in the opposite direction* (Fig. 4a). *If the triples of points A, B, C and C, D, A succeed each other in one direction, then the points B, C, D succeed each other in the same direction* (Fig. 4b). *For every pair of points A and B one can find points C and D such that the triples of points A, C, B and A, D, B succeed each other in opposite directions* (Fig. 4c). *The order of points is preserved under projection.* which means that if a line g is projected onto a line g' and if t is one of the directions on g, then g' can be provided with a direction t' such that whenever three points A, B, C on the line g succeed each other in the direction t, the corresponding points A', B', C' on the line g' succeed each other in the direction t'.

Before formulating the last axiom of this group, we note the following fact. If we fix any point of the line, then, by using the succession of triples of points containing the fixed point, we can introduce an order relation for pairs of points belonging to the rest

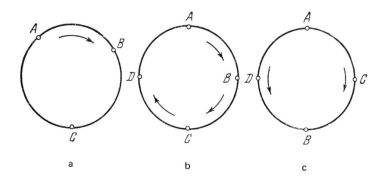

a   b   c

FIG. 4

of the line, and this relation will satisfy all the axioms of linear order except Dedekind's axiom. In fact, fixing a point C and a direction t of succession of three points on a line, we say that the point A precedes the point B if the three points A, B, C succeed each other in the direction t, and it is not hard to verify that this order relation satisfies the axioms in question. We now formulate the last axiom of the second group: *The order relation just introduced on the cut line (with the fixed point C omitted) must satisfy Dedekind's axiom.*

The most important sequence of the axioms of incidence and order in elliptic geometry is the possibility of introducing projective coordinates in elliptic space. In these coordinates, planes are specified by linear equations and lines by systems of two such equations.

The *third group* of axioms of elliptic geometry consists of the *axioms of congruence*. The axioms of this group will not be given here. Suffice it to say that the axioms of linear congruence and the axioms of order allow us to uniquely define the concept of the length of a segment (for a given unit of length) and that, just as in Euclidean geometry, this length is additive and has the same value on congruent segments.

§11. STATEMENT OF HILBERT'S PROBLEM

Let A_E denote the system of axioms of Euclidean geometry. Suppose that from the system A_E we drop all the axioms of congruence involving the concept of angle and add the following axiom to the resulting system: The length of any side of any triangle is less than the sum of the lengths of the other two sides (the triangle inequality). This gives a modification of the system of axioms A_E which we denote by A_E'. *Hilbert's Fourth Problem consists in determining, up to isomorphism, all realizations of the system of axioms A_E'.* We now explain what this means.

To *realize* a system of axioms means to specify objects of three kinds, called "points," "lines" and "planes," together with three relations between them, expressed by the words "belongs to," "precedes" and "is congruent to," for which the axioms of the system A_E' hold because of the concrete meaning of the objects and relations. This is illustrated by the following example.

Suppose we call an ordered triple (x_1, x_2, x_3) of real numbers a point and the numbers x_1, x_2, x_3 the coordinates of the point. Then we take a plane to mean the set of points whose coordinates satisfy a linear equation

$$a_1x_1 + a_2x_2 + a_3x_3 + a_0 = 0, \tag{1}$$

and this equation is called the equation of the plane. By a line we mean the intersection of two planes, provided that the intersection is nonempty. Thus a line is specified by a system

$$\left. \begin{array}{l} a_1x_1 + a_2x_2 + a_3x_3 + a_0 = 0, \\ a_1'x_1 + a_2'x_2 + a_3'x_3 + a_0' = 0 \end{array} \right\} \tag{2}$$

of two linear equations, and these equations are called the equations of the line. We say that a point belongs to a plane if its coordinates satisfy the equation of the plane. Correspondingly, we say that a point belongs to a line if its coordinates satisfy the equations of the line. Thus we have defined the concepts of point, line, and plane, together with an incidence relation for them. We now define an order relation, as expressed by the word "precedes."

As is well-known, the general solution of the system of equations (2) can be written in the form

$$x_i = \alpha_i + \beta_i t \quad (i = 1, 2, 3), \tag{3}$$

where the α_i, β_i are constants ($\beta_i \neq 0$) and t is a parameter. Let us agree to regard the point x' as preceding the point x'' in one of the two directions on the line if the parameter values corresponding to x' and x'' satisfy the inequality $t' < t''$, while x' precedes x'' in the other direction if $t' > t''$. The order relation defined in this way does not depend on the choice of the parameter t, since going over to another parametrization of the line involves a linear, and hence a monotonic, transformation of the parameter.

Finally we define a congruence relation for segments. Let $F(x_1, x_2, x_3)$ be an arbitrary positive, even, strictly convex function, which is positive homogeneous of degree 1. This means that

1) $F(x) > 0$ if $x \neq 0$,
2) $F(x) = F(-x)$,
3) $F(\lambda x) = |\lambda| F(x)$,
4) $F(\lambda x + \mu y) < \lambda F(x) + \mu F(y)$ $(\lambda, \mu > 0)$.

Then by the length of the segment with endpoints x and y we mean the quantity $|\overline{xy}| = F(x - y)$, and we say that two segments are congruent if they have the same length.

It is not hard to verify that all the axioms of the system A'_E are satisfied with these concrete meanings for the basic objects (point, line and plane) and for the relations of membership, order and congruence. For example, let us verify the validity of the axioms of congruence and of the "triangle inequality." First we note that the segments $\overline{xy}$ and $\overline{yx}$ are congruent, since the function F is even and hence $F(x - y) = F(y - x)$. The transivity of congruence of segments follows from the transitivity of the equality of real numbers. To verify the additivity, let x, y and z be three points on a line, where x precedes y and y precedes z in one of the two directions on the line. Then for these points, we have

$$x = \alpha + \beta t_1, \; y = \alpha + \beta t_2, \; z = \alpha + \beta t_3, \; t_1 < t_2 < t_3,$$
$$|\overline{xy}| = (t_2 - t_1) F(\beta),$$
$$|\overline{xz}| = (t_3 - t_1) F(\beta),$$
$$|\overline{yz}| = (t_3 - t_2) F(\beta).$$

It follows that

$$|\overline{xz}| = |\overline{xy}| + |\overline{yz}|.$$

To verify that the triangle inequality also holds, let x, y, z be the vertices of a triangle, so that

$$|\overline{xy}| = F(y - x), \qquad |\overline{xz}| = F(z - x), \qquad |\overline{yz}| = F(z - y).$$

Then

$$F(y - x) = F((y - z) + (z - x)) < F(y - z) + F(z - x),$$

i.e.,

$$|\overline{xy}| < |\overline{xz}| + |\overline{yz}|,$$

by the convexity and evenness of the function F.

The realization of the system of axioms A'_E just constructed is called the *Minkowski realization*, and the corresponding geometry is called *Minkowskian geometry*. Note that if the function F does not satisfy the convexity condition but satisfies the remaining three conditions, then we get a realization of the system of axioms A'_E without the "triangle inequality." The possibility of such a realization shows that the "triangle inequality" cannot be proved by using the remaining axioms of the system A'_E.

The Minkowski realization has a simple intuitive interpretation. Suppose that in Euclidean space we introduce a central symmetric convex surface Φ with its center of symmetry at the origin of coordinates. We will measure an arbitrary segment by the diameter of the surface Φ parallel to the segment, defining congruence of segments to be equality of their lengths. The validity of the congruence axioms of the system A'_E can then be readily confirmed. As for the validity of the remaining axioms of A'_E, they hold trivially, since the space in which the realization is constructed is Euclidean. If the diameter of the surface Φ parallel to the vector x is denoted by $d(x)$, then the function $F(x)$ used to define the length of a segment in the Minkowski realization is related to $d(x)$ by the simple formula

$$F(x) = \frac{\|x\|}{d(x)},$$

where $\|x\| = (x_1^2 + x_2^2 + x_3^2)^{\frac{1}{2}}$ is the ordinary Euclidean length of x. We note that the Euclidean realization is obtained in the case where the surface Φ is an ellipsoid.

We now explain the concept of an isomorphism of two realizations. Two realizations R_1 and R_2 of the system of axioms A'_E are said to be *isomorphic* if a one-to-one correspondence can be established between the points, lines and planes of the realizations which preserves the relations of incidence, order and congruence. This means the following: If α_1 is a plane and A_1 is a point in the realization R_1, while α_2 and A_2 are the corresponding plane and point in the realization R_2, then A_2 belongs to α_2 if and only if A_1 belongs to α_1. If a_1 is a line in the realization R_1 and t_1 is any direction on a_1, then a direction t_2 can be specified on the corresponding line a_2 in the realization R_2 such that whenever $A_1 < B_1$ for two points A_1 and B_1 on the line a_1, the corresponding points A_2 and B_2 on the line a_2 in the realization R_2 stand in the same order relation, i.e., $A_2 < B_2$. If two segments A_1B_1 and C_1D_1 are congruent in the realization R_1, then the corresponding segments A_2B_2 and C_2D_2 in the realization R_2 are also congruent.

As we now show, there are nonisomorphic realizations among the Minkowski realizations of the axiom system A'_E. To this end, we first establish a condition under which two Minkowski realizations, defined by using the surfaces Φ_1 and Φ_2, are isomorphic. If the realizations are isomorphic, then the Euclidean space in which the realizations are constructed has a mapping onto itself carrying lines into lines, planes into planes, and congruent segments into congruent segments. This mapping must be projective, since it carries planes into planes and lines into lines. Moreover, it is affine, since it maps the whole space into itself.

Let $\mu_1(\delta)$ be the length of a segment δ in the first realization, and let $\mu_2(\delta)$ be the length of the segment in the second realization, corresponding to δ under the isomorphism. The quantity $\mu_2(\delta)$, regarded as a function defined on segments of the first realization, has all the properties of length, i.e., it is positive and additive, and it takes the same value on congruent segments. Therefore $\mu_1(\delta) = k\mu_2(\delta)$, since the length of a segment is uniquely determined to within a constant factor depending on the choice of the unit of length. We can assume without loss of generality that $\mu_1(\delta) = \mu_2(\delta)$. This means that corresponding segments in the two realizations have the same length. It follows that the affine transformation establishing the isomorphism carries the surface Φ_2 into the surface Φ_1. Thus, for the realizations to be isomorphic, it is necessary that the surfaces Φ_1 and Φ_2 specifying the realizations be affinely equivalent, i.e., that they be carried into each other by an affine transformation. This condition is obviously sufficient, i.e., an affine transformation of space, carrying the surface Φ_2 into Φ_1, gives an isomorphic mapping of one realization onto the other.

As a rule, two arbitrary central symmetric convex surfaces are not affinely equivalent and the corresponding Minkowski realizations are, therefore, not isomorphic. To give a concrete example of affinely nonequivalent surfaces, let Φ_1 be an ellipsoid and let Φ_2 be any surface other than an ellipsoid. Then these surfaces are obviously not affinely equivalent, since any surface affinely equivalent to an ellipsoid is itself ellipsoid.

We have explained the statement of Hilbert's problem, starting from the system of axioms of Euclidean geometry. There is an analogous statement of the problem for the case of Lobachevskian geometry. In fact, suppose that from the system of axioms for Lobachevskian geometry A_L we drop all the axioms of congruence involving the concept of angle and add the "triangle inequality" axiom to the resulting system. This gives a modification of the system of axioms A_L which we denote by A'_L. *Hilbert's fourth problem consists in determining up to isomorphism all realizations of the system of axioms A'_L.*

A large class of realizations of the system of axioms A'_L was given by Hilbert himself [14]. These realizations are obtained as follows. Let G be any bounded strictly convex domain in Euclidean space. Then the points in Hilbert's realization are taken to be the points of the domain G, while the lines and planes are taken to be the intersections of Euclidean lines and planes with the domain G. The order of points on the line is understood in the sense of Euclidean geometry. Two segments are said to be congruent if their lengths are equal, and the length $d(x, y)$ of a segment $\overline{xy}$ is defined by the formula

$$ d(x, y) = k \left| \ln \left(\frac{d_0(x, u)}{d_0(x, v)} : \frac{d_0(y, u)}{d_0(y, v)} \right) \right| . $$

Here u and v are the points of intersection of the Euclidean line xy with the boundary of the domain G, and d_0 denotes the Euclidean length of the segment with the given endpoints. If the domain G is chosen to be an ellipsoid, we obtain the well-known Cayley-Klein realization of the system of axioms of Lobachevskian geometry. In Hilbert's realization all the axioms of the system A'_L other than the "triangle inequality" are satisfied in an obvious way. Moreover, the "triangle inequality" is also satisfied, as we now show.

Take an arbitrary triangle with vertices x, y, z (Fig. 5). Since the cross ratio of four points on a line is preserved under projection, we have

$$ d(x, z) = k \left| \ln \left(\frac{d_0(x, u')}{d_0(x, v')} : \frac{d_0(w, u')}{d_0(w, v')} \right) \right| , $$

$$ d(y, z) = k \left| \ln \left(\frac{d_0(w, u')}{d_0(w, v')} : \frac{d_0(y, u')}{d_0(y, v')} \right) \right| , $$

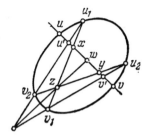

FIG. 5

and hence

$$d(x, z) + d(y, z) = k\left|\ln\left(\frac{d_0(x, u')}{d_0(x, v')} : \frac{d_0(y, u')}{d_0(y, v')}\right)\right|.$$

If we replace u' and v' by u and v in the right-hand side of this formula, then it obviously becomes smaller. On the other hand, the right-hand side goes into $d(x, y)$ under this change of variables. Therefore

$$d(x, y) < d(x, z) + d(y, z),$$

as was to be proved.

Just as in the case of Minkowski realizations of the system of axioms A_E', it can be shown that two Hilbert realizations of the system of axioms A_L', defined in domains G_1 and G_2, are isomorphic if and only if the domains G_1 and G_2 are projectively equivalent, i.e., are carried into each other by a projective transformation. Hence if the domains G_1 and G_2 are projectively nonequivalent, the Hilbert realizations defined in them are nonisomorphic. In particular, we get nonisomorphic realizations if we take G_1 to be the domain bounded by an ellipsoid and G_2 to be the domain bounded by any non-ellipsoidal surface.

Finally we consider the formulation of Hilbert's problem in the elliptic case. Let A_R denote the system of axioms of elliptic geometry, and suppose that in the system A_R we drop all the axioms of congruence involving the concept of angle and add the "triangle inequality" axiom to the resulting system. This gives a modification of the system of axioms A_R which we denote by A'_R. *Hilbert's fourth problem consists in determining to within an isomorphism all realizations of the system of axioms A'_R.*

REMARK. In the elliptic case, the "triangle inequality" needs further clarification. The point is that in elliptic geometry a pair of points x, y determines two segments on the line xy. One segment is defined as the set of points z for which the three points x, z, y succeed each other in one direction, and the other segment is defined as the set of points z for which the points x, z, y succeed each other in the opposite direction. Every point z of the line different from x and y belongs to one of these segments. The points x and y themselves are called the endpoints of the segments. By a triangle we mean a figure which consists of three noncollinear points and three segments that join these points in pairs and do not intersect a plane. It is not hard to see that the Axiom of Pasch holds with this definition of the concept of a triangle. The "triangle inequality" holds for triangles in the sense of this definition. (Compare Note (5).)

§12. SOLUTION OF HILBERT'S PROBLEM

In this section we shall solve Hilbert's fourth problem, in the form in which it was posed in the preceding section. It will be recalled that we are talking about determining, up to isomorphism, all realizations of the axiom systems A'_E, A'_L and A'_R, which are obtained from the axioms of Euclidean geometry, Lobachevskian geometry and elliptic geometry if we drop all the axioms of congruence involving the concept of angle and add the "triangle inequality" axiom to the resulting systems. First we consider the two-dimensional case of the problem. In this case, the spatial axioms of membership are dropped in the systems A'_E, A'_L and A'_R, and the resulting systems are supplemented with Desargues' theorem, as an axiom. The corresponding systems of axioms in the two-dimensional case will be denoted by A_E'', A_L'' and A_R''.

Let α be a Euclidean plane, and fix a point O in α. We complete the plane α by adjoining ideal elements, and we denote the completed (and now projective) plane by $\bar{\alpha}$. In the plane $\bar{\alpha}$ we define a function σ which is completely additive on Borel sets and satisfies the following conditions:

1) $\sigma(X) \geqslant 0$ for every set X;
2) $\sigma(X) = 0$ if X lies on a line;
3) $\sigma(X) > 0$ if X is an open set;
4) $\sigma(X) < \infty$ outside every neighborhood of the point O;
5) $\sigma(X) = \infty$ on every half-plane α_g of the plane α whose boundary g passes through the point O.

Next we introduce homogeneous coordinates x_1, x_2, x_3 in the plane $\bar{\alpha}$ in such a way that the point O has coordinates $0, 0, 1$, and the ideal line (i.e., the line at infinity) has the equation

$$x_3 = 0.$$

Let H denote the correlation of the plane $\bar{\alpha}$ under which the point with homogeneous coordinates a_1, a_2, a_3 is assigned the line with the equation

$$a_1x_1 + a_2x_2 + a_3x_3 = 0.$$

We define a complete additive function σ' on the set of lines of the plane $\bar{\alpha}$ by setting

$$\sigma'(G) = \sigma(H^{-1}\,G),$$

for an arbitrary set G, where $H^{-1}G$ is the pre-image of the set of lines G under the correlation H.

We now construct a realization S_E'' of the system of axioms A_E'' as follows. We take the points of S_E'' to be the points of the Euclidean plane α and the lines of S_E'' to be the lines of α. The relations of incidence and order are understood in the sense of Euclidean geometry. Finally, two segments are defined as being congruent if their lengths are equal, and the length of a segment is defined as the value of the function σ' on the set of lines intersecting the segment.

The first two groups of axioms of the system A_E'' are satisfied in an obvious way in the realization S_E'', since they are satisfied in Euclidean geometry. It is not hard to see that the remaining axioms of the system A_E'' also hold in S_E''. In fact, it is obvious that the segment AB is congruent to the segment BA. If is also obvious that if AB is congruent to the segments CD and EF, then CD and EF are themselves congruent. Moreover, if B and B_1 are points of the segments AC and A_1C_1, respectively, then the congruences $AB = A_1B_1$ and $BC = B_1C_1$ imply the congruence $AC = A_1C_1$. The validity of this axiom follows from the additivity of the function σ and condition 2) satisfied by σ. The possibility of laying off a line segment from a given point on a line in a given direction, which is congruent to a given segment, follows from the complete additivity of the function σ and condition 5). Finally, the "triangle inequality" follows from the additivity of the function σ and this Axiom of Pasch. A realization S_E'' constructed in this way will be called a *σ-realization* of the system of axioms A_E''.

We now construct a σ-realization of the system of axioms A_L''. To this end, in the plane α we choose any bounded convex domain B containing the point O. Under the correlation H, a set of lines intersecting the domain B is assigned a set of points lying outside some convex domain B'. Outside B' we define a completely additive function σ satisfying conditions 1)-3), and we replace conditions 4) and 5) by the following conditions:

4') $\sigma(X) < \infty$ outside every neighborhood of the closure of the domain B';

5') $\sigma(X) = \infty$ on every half-plane α_g bounded by a supporting line g of the domain B' (we are talking about the half-plane which does not contain the domain B').

The realization S_L'' is constructed as follows. We take the points of S_L'' to be the points of the domain B and the lines of S_L'' to be the intersections of the Euclidean lines of the plane α with the domain B. The relations of incidence and order are understood in the sense of Euclidean geometry in the plane α. Two segments are said to be congruent if their lengths, defined with the help of the function σ', are equal. The validity of the axioms in the realization S''_L is verified in the same way as in the realization S''_E.

A σ-realization S''_R of the system of axioms A''_R is constructed similarly. In this case the completely additive function σ is defined in the whole plane $\bar{\alpha}$. The function σ is subjected only to the conditions 1)-3) and the condition of being bounded in the whole plane $\bar{\alpha}$. We take the points of the realization S''_R to be the points of the projective plane $\bar{\alpha}$ and the lines of S''_R to be the projective lines. The relations of incidence and order are understood in the sense of projective geometry, and congruence of segments is defined with the help of the function σ'.

Theorem 1. *Every realization of any of the axiom systems A''_E, A''_L and A''_R is isomorphic to a σ-realization.*

PROOF. Let S be an arbitrary realization of the system of axioms A''_R. The first two groups of axioms of the system A''_R (the axioms of incidence and order) constitute the axiom system of projective geometry in the plane. It follows that the realization S of the axiom system A''_R is at the same time a realization of the axiom system of projective goemetry in the plane. As is well known, the axiom system of projective geometry is categorical. This means all of its realizations are isomorphic. It follows that we can establish a one-to-one correspondence between the points and lines of the realization S and the points and lines of the projective plane $\bar{\alpha}$ which preserves the relations of membership and order. To define the relation of congruence for segments in the plane $\bar{\alpha}$, we now say that two segments in the plane $\bar{\alpha}$ are congruent if the segments corresponding to them in the realization S are congruent. It is obvious that with this interpretation of congruence of segments, we obtain in the projective plane $\bar{\alpha}$ a realization $S_{\bar{\alpha}}$ of the axiom system A''_R, such that $S_{\bar{\alpha}}$ is isomorphic to S. We now show that the realization $S_{\bar{\alpha}}$ is a σ-realization in the sense of the definition given above.

Using the axioms of order and the axioms of congruence in the system A''_R, which hold in the realization $S_{\bar{\alpha}}$, we can uniquely determine (up to a constant factor) the concept of the length of a segment in the plane $\bar{\alpha}$, which has the usual properties (positivity, additivity, equality on congruent segments). Having defined the length of segments, we metrize the plane $\bar{\alpha}$, by calling the smaller of the lengths of the two segments joining two given points x and y the distance between x and y. This metrization is obviously Desarguesian (the geodesics are straight lines). Two segments in the realization $S_{\bar{\alpha}}$ are congruent if and only if they have the same length in the Desarguesian metric just constructed. Since in the two-dimensional case, every Desarguesian metric is a σ-metric, the realization $S_{\bar{\alpha}}$ is a σ-realization.

Next let S be an arbitrary realization of the system of axioms A_E''. We begin by introducing homogeneous coordinates in the realization S, just as is done in the projective plane. We first introduce such coordinates inside an arbitrary triangle. As is well-known, the introduction of homogeneous coordinates in the projective plane is based in an essential way on Desargues' theorem and harmonic conjugacy of points (see [5]), Desargues' theorem is introduced as an axiom in the system of axioms A_E'', with certain stipulations on the existence of the points of intersection figuring in the theorem. But these stipulations cause no trouble in introducing homogeneous coordinates inside a triangle, provided that the unit point with coordinates 1, 1, 1 lies inside the triangle.

Once homogeneous coordinates have been introduced inside an arbitrary triangle Δ_0, they can be extended to any point of the realization S in the following way. Let x_i be the homogeneous coordinates inside Δ_0, and take an arbitrary point x. We construct an arbitrary triangle Δ which contains the point x and overlaps the triangle Δ_0. In the triangle Δ we introduce homogeneous coordinates, which we denote by x_i'. Consider a mapping of the projective plane, under which the point with coordinates x_i is assigned the point with coordinates x_i, where x_i' and x_i are the coordinates of one and the same point belonging to both triangles Δ_0 and Δ. This mapping carries lines into lines, and hence is linear. It follows that $x_i = \sum_j A_i^j x_j'$ on the intersection of the triangles Δ_0 and Δ. We use this formula to extend the system of coordinates from Δ_0 to the triangle

Δ, in particular to the point x. It is obvious that this method of extending the coordinates to the point x does not depend on the choice of the triangle or on the system of coordinates (x_i') inside Δ.

Having constructed homogeneous coordinates in S, we map S onto the projective plane $\bar{\alpha}$, by assigning to an arbitrary point x the point of the plane $\bar{\alpha}$ with the same homogeneous coordinates. Since Euclid's axiom of parallels holds in S, it is not hard to see that the image of S under this mapping is the whole plane $\bar{\alpha}$ cut along some line. Let α denote the plane $\bar{\alpha}$ cut in this way. We now define the concept of congruence for segments of the plane α, by regarding two segments as congruent if the segments corresponding to them in the realization S are congruent. With this definition of congruence of segments, we obtain in the plane α a realization S_α of the axiom system A_E'' which is isomorphic to S. Moreover, as in the elliptic case considered above, we conclude that S_α is a σ-realization.

The argument is similar in the case of a realization S of the system of axioms A_L''. The only difference is that the image of S in the prospective plane $\bar{\alpha}$ will be a convex domain, whose closure does not intersect some line. This follows from the Lobachevskian axiom of parallels, which is contained in A_L''.

We now turn to the three-dimensional case of Hilbert's problem. We take Euclidean space, and in it we define an arbitrary complete σ-metric in the generalized sense (Sec. 9). We then define a congruence relation for segments, by saying that two segments are congruent if their lengths are equal in this metric. This gives a realization of the system of axioms A_E', which we call a σ-realization in the generalized sense. Similarly, we define a σ-realization in the generalized sense for the system of axioms A_L'. The only difference is that the domain of definition of the metric will be a bounded convex domain, and the realization will be defined in this domain. In the case of the system of axioms A_R', we take projective space instead of Euclidean space, and define a generalized σ-metric in it.

Theorem 2. *Every realization of any of the axiom systems A_E', A_L' and A_R' is isomorphic to a σ-realization in the generalized sense.*

The proof of Theorem 2 is essentially a repetition of the proof of Theorem 1.

NOTES

[1] By a domain, here, is meant a convex open set.

[2] For a recent comprehensive account of the significance and the status of Hilbert's Fourth Problem, including a description of the present work, see Busemann [8].

[3] Metrics for which the geodesics are straight lines are also called projective metrics.

[4] More precisely, we begin with the set of non-zero vectors of R^4 and define two vectors $x = (x_1, x_2, x_3, x_4)$ and $y = (y_1, y_2, y_3, y_4)$ to be equivalent, $x \sim y$ in symbols, if there is a non-zero number λ such that $y = \lambda x$. The relation $\sim$ is easily seen to be an equivalence relation. The projective space P^3 is defined to be the quotient set of $R^4 \setminus \{0\}$ by the relation $\sim$. See Busemann and Kelly [9] for a more detailed treatment in a rather similar spirit.

[5] It is easy to appreciate the intuitive significance of the definition of projective triangle if we consider the projective plane to be realized as the unit sphere with antipodal pairs of points identified. Consider three nearby points x^1, x^2, and x^3. Join x^1 to x^2 and x^3 with the shorter of the two projective line intervals joining those points. Join x^2 to x^3 by the longer of the two projective line

intervals joining those points. Then there are lines meeting the side $x^2 x^3$ which avoid both of the remaining sides of the triangle.

[6] The reader who is uncomfortable with arguments involving the renormalization of homogeneous coordinates may prefer the treatment of this result given in [9], especially (I.6).

[7] See Busemann [5] for a complete account of the concepts of the length of curves in a metric space, metric segments, and geodesics, especially Chapter I, Sec. 5-7. [9] gives a somewhat more elementary account in which, however, Desarguesian metrics are called projective metrics.

[8] Royden [20] is a good reference on topological spaces, completely additive functions, Borel sets, and related notions.

[9] See Busemann [5], pp. 82-84, for a more precise statement as well as a discussion of the geometric significance of the assumptions.

[10] Although the proof of the Theorem is self-contained, it does not illuminate the geometric significance of the two expressions for ds and $d\sigma$, involved in its statement. We wish to explain this as it is the essence of Busemann's beautiful idea from integral geometry which, as already explained in the Preface, was the starting point for this investigation.

A line in the euclidean plane may be specified by a pair of numbers (p, φ), its so-called *normal coordinates,* in terms of which the equation of the line takes the form

$$x_1 \cos \varphi + x_2 \sin \varphi = p . \tag{1}$$

A set S of straight lines in the plane may be regarded as the set of the corresponding pairs of normal coordinates (p, φ), which we shall also denote by S. We define the set function σ, on the set of lines in the plane, by

$$\sigma (S) = \iint_S \gamma \, dp \wedge d\varphi , \tag{2}$$

where γ is a constant. It can be shown (see Santalo [21], [22], and [23], pp. 147-193) that this set function σ is, up to the arbitrary constant γ, the only measure on the sets of lines in the plane. By a measure, we mean a completely additive set function which is given in this way as an integral and which is invariant under the euclidean motions. The differential 2-form $d\sigma = \gamma\, dp \wedge d\varphi$, which occurs under the integral sign in (2), is called the *density* for straight lines. It is always taken in absolute value since we desire a nonnegative measure. It is precisely this density which, through a slight abuse of notation, is given as $d\sigma = \gamma\, dp\, d\xi$ in the statement of the Theorem.

Next, consider a fixed line g. We wish to calculate the length of an interval of g, with the help of our set function σ, as it was defined in Section 3. Denote by $\overline{dx}$ the interval of g from x to $x + dx$, where x is a point of g and dx is a vector parallel to g. We obtain

$$|\overline{dx}| = \sigma(\pi\overline{dx}) = \int_{p=x}^{x\,\cdot\,\xi + dx\,\cdot\,\xi} \int_{\varphi=0}^{\pi} \gamma\, dp \wedge d\varphi =$$

$$= \int_{\varphi=0}^{\pi} \gamma\, (dx \cdot \xi)\, d\varphi \tag{3}$$

It is precisely this expression which, through a slight abuse of notation, is given as

$$ds = \int_{\omega} \gamma\, (g)\, |\xi\, dx|\, d\xi$$

in the statement of the Theorem.

To reconcile the notations, put $\xi = (\cos\,\varphi,\, \sin\,\varphi)$ so that the normal form of the equation of a line becomes $\xi \cdot x = p$. We then have $d\xi = (-\sin\,\varphi,\, \cos\,\varphi)\, d\varphi$ so that $|d\varphi| = \|d\xi\|$.

The class of σ-metrics obtained in this way does not exhaust the class of regular Finsler metrics. In fact, in applying (3), we must allow the constant γ to vary with the line g. Equivalently, we may regard $\gamma = \gamma\,(x,\, dx)$ as depending on a point and a direction at the point so that ds is evidently of the form $F\,(x,\, dx)$.

[11] The Euler equations are derived in books on the Calculus of Variations, see Akhiezer [1] and Gelfand and Fomin [11], especially Chapter 2, Section 10.

[12] Equations (1) may be viewed in either of two ways. If we suppose that F is given, then (1) is a system of ordinary differential equations for the geodesics. On the other hand, if we suppose that the geodesics are given, then (1) is a system of partial differential equations which are satisfied by the function F which specifies the metric. In his original solution of Hilbert's Fourth Problem, referred to in the Introduction, Hamel [12] obtained the solutions of (1) for F. However, as the standard theory of partial differential equations only yields smooth solutions, Hamel was only able to obtain the regular Desarguesian metrics in this way.

[13] Implicit in this discussion is the elementary fact that any subsegment of a metric segment is again a metric segment, see [5], (6.2).

[14] The method being used here to smooth the only continuous metric $\rho(x, y)$, to obtain the smooth metric $\rho_\delta(x, y)$, is essentially convolution of $\rho(x, y)$ with the "bump" function $\varphi(r)$. Since $\varphi(r)$ is zero outside the ball $\|r\| < \delta$, the integration may equally well be considered to be extended over the entire space. Introducing the change of variables $q = z + r$ and, therefore, $dq = dr$, yields

$$\int \varphi(r)\, \rho(x + z + r,\ y + z + r)\ dr = \int \varphi(q - z)\, \rho(x + q,\ y + q)\ dq$$

from which the claimed continuous differentiability is apparent.

[15] A *convex body* is a bounded closed convex set.

[16] The systems of axioms for the classical geometries, which are given here, follow Pogorelov [19] rather than Hilbert [14]. Section 10 is a terse summary of the relevant parts of [19], the concept of congruence being treated somewhat differently.

BIBLIOGRAPHY

1. Akhiezer, N. I., "The Calculus of Variations," Blaisdell, New York, 1962.
2. Blaschke, W., "Kreis und Kugel," 2nd ed., de Gruyter, Berlin, 1956.
3. Blaschke, W. and Leichtweiss, K., "Elementare Differential-geometrie," 5th ed., Springer, Berlin and New York, 1973.
4. Bonnesen, T. and Fenchel, W., "Theorie der konvexen Körper," Springer, Berlin, 1934.
5. Busemann, H., "Geometry of Geodesics," Academic Press, New York, 1955.
6. Busemann, H., "Areas in Affine Spaces III: The Integral Geometry of Affine Area," *Rend. Circ. Mat. Palermo* **9** (1960) 226–242.
7. Busemann, H., "Geometries in which Planes Minimize Area," *Annali Mat. Pure Appl. (IV)* **55** (1961) 171–190.

8. Busemann, H., "PROBLEM IV: Desarguesian Spaces," *Proceedings of Symposia in Pure Mathematics* [American Mathematical Society], 28 (1976) 131–141.

9. Busemann, H. and Kelly, P., "Projective Geometry and Projective Metrics," Academic Press, New York, 1953.

10. Darboux, G., "Leçons sur la théorie générale des surfaces, III Partie," Gautier-Villars, Paris, 1894.

11. Gelfand, I. M. and Fomin, S. V., "Calculus of Variations," Prentice-Hall, Englewood Cliffs, 1963.

12. Hamel, G., "Über die Geometrieen, in denen die Geraden die Kürzesten sind," *Math. Ann.* 57 (1903) 232–264.

13. Hilbert, D., "Mathematical Problems," Lecture delivered before the International Congress of Mathematicians at Paris in 1900, trans. M. W. Newson, *Bull. Amer. Math. Soc.,* 8 (1902) 437–479.

14. Hilbert, D., "Foundations of Geometry," Translated from the 10th ed., Open Court, Lasalle, 1971.

15. Hilbert, D., "Über die gerade Linie als kürzeste Verbindung zweier Punkte," *Math. Ann.,* 46 (1895) 91–96.

16. Hirsch, A., "Über eine characteristische Eigenschaft der Differentialgleichungen der Variationsrechnung," *Math. Ann.,* 49 (1897) 49–72.

17. Minkowski, H., "Geometrie der Zahlen," Leipzig and Berlin, 1910.

18. Pogorelov, A. V., "A Complete Solution of Hilbert's Fourth Problem," *Sov. Math.,* 14 (1973) 46–49.

19. Pogorelov, A. V., "Lectures on the Foundations of Geometry," Noordhoff, Groningen, 1966.

20. Royden, H., "Real Analysis, 2nd ed., Macmillan, New York, 1968.

21. Santalo, L. A., "Introduction to Integral Geometry," Hermann, Paris, 1953.

22. Santalo, L. A., "Integral Geometry," in Chern, S. -S., "Studies in Global Geometry and Analysis," Prentice-Hall, New York, 1967.

23. Santalo, L. A., "Integral Geometry and Geometric Probability," Addison-Wesley, Reading, 1976.

INDEX

Akhiezer, N. I., 90
axioms of classical geometries, 68

Borel sets, 21
Busemann, H., 20, 24, 88, 89

Cayley–Klein realization, 80
Christoffel's theorem, 61
closed sets, 20
complete quadrangle, 72
completely additive set functions, 20, 21
congruence, 70
convex body, 91

Darboux, G., 7
Dedekind's axiom, 69, 74
density, 89
Desarguesian metrics, 3, 19, 20
regular approximation of, 38
Desarguesian spaces, 7
Diagonal points, 72
distance, 19, 25
division of pairs of points, 14

elliptic geometry, see
Riemannian geometry, 9
elliptic type metric, 39
equivalent quadruples, 9
Euclidean geometry, 7, 72, 79, 82
Euler equations, 26, 62, 90

Finsler metrics, 25, 90
Funk's problem, 54, 55

Gelfand, I. M., 90
geodesics, 3

half-plane, 69
Hamel, G., 2, 3, 7, 90
harmonic conjugacy, 72
harmonic division, 14
Hilbert realizations, 80
Hilbert's fourth problem,
    statement of 5, 75
    solution of 82
homogeneous coordinates, 9
hyperbolic geometry, see
        Lobachevskian geometry
hyperbolic type metric, 38

incidence, axioms of, 68
interval, 12
isomorphism of realizations, 78,
    85

Laplace operator, 58
line, 9
Lobachevskian geometry, 5, 7,
    71, 72, 79, 80, 82

metrics 19
Minkowski, H., 6
Minkowski realization, 77, 78
Minkowski's geometry, 7, 77

neighborhood of a point, 20
nonhomogeneous coordinates,
    24
normal coordinates, 89

open sets, 20
opposite sides, 72
order, axioms of, 68

parabolic type metric, 39
parallels, axiom of, 68
Pasch, axiom of, 12, 223, 23,
    70, 82
plane, 9
point, 9, 13
projective mappings, 13
projective metrics, 88
projective space, 9
projective transformations, 13
projective triangle, 12

quadrangle, 72

realization of a system of
    axioms, 75
regular Desarguesian metrics,
    24
regular metric, 25
Riemannian geometry, 5, 7,
    72
Royden, H., 88

segment, 20, 69

sides of a triangle, 12, 71
σ-metric, 23
   in the generalized sense, 67
σ-realization, 84
support function, 54

three-dimensional case of
   Desarguesian metric, 61

transformations, projective, 13
triangle inequality, 19, 22, 76,
   80, 82
two-dimensional case of
   Desarguesian metric, 46

Veronese, 6
vertices of a triangle, 12,
   70